T0138772

Understanding Nanoelectromechanical Quantum Circuits and Systems (NEMX) for the Internet of Things (IoT) Era

RIVER PUBLISHERS SERIES IN ELECTRONIC MATERIALS AND DEVICES

Indexing: All books published in this series are submitted to the Web of Science Book Citation Index (BkCI), to SCOPUS, to CrossRef and to Google Scholar for evaluation and indexing.

The "River Publishers Series in Electronic Materials and Devices" is a series of comprehensive academic and professional books which focus on the theory and applications of advanced electronic materials and devices. The series focuses on topics ranging from the theory, modeling, devices, performance and reliability of electron and ion integrated circuit devices and interconnects, insulators, metals, organic materials, micro-plasmas, semiconductors, quantum-effect structures, vacuum devices, and emerging materials. Applications of devices in biomedical electronics, computation, communications, displays, MEMS, imaging, micro-actuators, nanoelectronics, optoelectronics, photovoltaics, power ICs and micro-sensors are also covered.

Books published in the series include research monographs, edited volumes, handbooks and textbooks. The books provide professionals, researchers, educators, and advanced students in the field with an invaluable insight into the latest research and developments.

Topics covered in the series include, but are by no means restricted to the following:

- Integrated circuit devices
- Interconnects
- Insulators
- Organic materials
- Semiconductors
- Qantum-effect structures
- Vacuum devices
- Biomedical electronics
- Displays and imaging
- MEMS
- Sensors and actuators
- Nanoelectronics
- Optoelectronics
- Photovoltaics
- Power ICs

For a list of other books in this series, visit www.riverpublishers.com

Understanding Nanoelectromechanical Quantum Circuits and Systems (NEMX) for the Internet of Things (IoT) Era

Héctor J. De Los Santos

NanoMEMS Research, LLC
USA

River Publishers

Published, sold and distributed by:
River Publishers
Alsbjergvej 10
9260 Gistrup
Denmark

River Publishers
Lange Geer 44
2611 PW Delft
The Netherlands

Tel.: +4536995197
www.riverpublishers.com

ISBN: 978-87-7022-128-3 (Hardback)
978-87-7022-127-6 (Ebook)

Este libro lo dedico a mis queridos padres
y a mis queridos Violeta, Mara, Hectorcito,
y Joseph.

I dedicate this book to my dear parents
and to my dear Violeta, Mara, Hectorcito,
and Joseph.

Héctor J. De Los Santos

"Y sabemos que a los que aman a Dios todas las cosas les ayudan abien, esto es, a los que conforme a su propósito son llamados."

"And we know that all things work together for good to them that love God, to them who are the called according to his purpose."

Romanos 8:28

Contents

Preface

This book presents a unified exposition of the physical principles at the heart of nanoelectromechanical quantum circuits and systems (NEMX) in the context of understanding its potential applications to enable the Internet of Things era. The book delves into physical phenomena that permeate the operation of NEMX which, by definition, exploit nanoscale mechanical devices and novel nanoscale materials. These devices are expected, in turn, to be crucial to enable the realization of the IoT, which will be predicated upon the wireless connection of smart nodes consisting of sensors, actuators, energy harvesting and storage components, and wireless communications transceivers.

This book contains six chapters. Chapter 1 provides an introduction to the IoT, including its origins, its predicted impact on society, both in terms of improving its quality of life, and also its potential to permeate all spheres of society, from health care, to agriculture, to industry, to commerce, to the infrastructure, to the economy, and many more. Chapter 2 presents a brief introduction to the technical areas of microelectromechanical systems (MEMS) and nanoelectromechanical systems (NEMS), including their origins, their impetus and motivation for their development, and the plethora of physical phenomena that they bring to our disposal as tools for enabling one to engineer systems for solving a variety of problems. Chapter 3 engages in developing a comprehensive understanding of MEMS/NEMS physics, as would be pertinent for attaining the technical understanding to master the application and invention of these devices. In particular, we discuss the following topics: Actuation, Electrostatic Actuation, the Parallel-Plate Capacitor, the Electrostatically Actuated Cantilever Beam, the Interdigitated (Comb-drive) Capacitor, Piezoelectric Actuation, the Piezoelectric Cantilever Probe, Casimir Actuation, Casimir's Force Calculation Method, Lifshitz's Calculation of the Casimir Force, the Casimir Force Calculation of Brown and Maclay, Casimir Force Calculations for Arbitrary Geometries, Computing the Casimir Energy Based on Multipole Interactions, Computing the Casimir Force Using Finite-Difference Time-Domain Techniques, Computing the

Casimir Force Using the Framework of Macroscopic Quantum Electrodynamics, Corrections to Ideal Casimir Force Derivation, Radiation Pressure Actuation, Radiation Pressure Manipulation of Particles, Radiation Pressure Trapping of Particles, and the Radiation Pressure Effect on Cantilever Beams. Then, the topic of Mechanical Vibration, as it applies to single- and many-degrees-of-freedom systems, and Rayleigh's Method for calculating mechanical resonance frequency, is discussed. This is followed by an in-depth treatment of Thermal Noise in MEMS/NEMS, in particular, the Fundamental Origin of Intrinsic Noise in mechanical structures, and the Amplitude of Brownian (Random) Displacement of Cantilever Beam. We conclude with the topics of Sensing, The Accelerometer, Capacitive Accelerometer Implementation, Quantum Mechanical Tunneling Accelerometer, and Vibration Sensors. Chapter 4 deals with MEMS/NEMS Switches, Nanoelectromechanical Switches, including Downscaled MEMS/NEMS Switches and MEMS/NEMS Switches via Novel Materials, MEMS/NEMS Varactors, Nanoelectromechanical Varactors, including Dual-Gap MEMS/NEMS Varactors and MEMS/NEMS Varactors via New Materials, MEMS/NEMS Resonators, Nanoelectromechanical Resonators, including Clamp–Clamp RF MEMS Resonators and MEMS/ NEMS Resonators via New Materials. Chapter 5 addresses the topic of Understanding MEMS/NEMS for Energy Harvesting, including an introduction to Wireless Energy Harvesting, including the RF-DC Conversion Circuit and Resonant Amplification of Extremely Small Signals, and Mechanical Energy Harvesting, including Theory of Energy Harvesting from Vibrations, Piezoelectric Conversion, and Electrostatic Conversion. Chapter 6 exposes NEMX Applications in the IoT Era. We begin with an introduction to the fundamentals of IoT networks and nodes, including, Wireless Connectivity, Communication Protocols, Network Range, and The Origins of the IoT, and we expose how NEMX components are encroaching in a variety of IoT applications such as the Smart Home, Wearables, the Smart City, the Smart Grid, the Industrial Internet; the Connected Car, Connected Health, Smart Retail, the Smart Supply Chain, and Smart Farming. Then, we address NEMX Applications in Wireless Sensor Networks, in particular applications in radios and in agriculture. Finally, we address the subject of the IoT in the context of emerging mmWaves/5G (fifth-generation wireless networks) technology. In particular, we expose 5G from the Systems and Technologies viewpoints, the motivation for exploiting the advantages afforded by mmWaves/5G in the context of the IoT and its potential impact via the enablement of Device-to-Device Communications, and Simultaneous Transmission/Reception, and Potential mmWave/5G

Frequencies for the IoT. The book ends with two appendices, namely Appendix A, MEMS Fabrication Techniques Fundamentals, and Appendix B, Emerging Fabrication Technologies for the IoT: Flexible Electronics and Printed Electronics.

The book assumes a preparation at the advanced undergraduate/beginning graduate student level in Physics, Electrical Engineering, Materials Science, or Mechanical Engineering. It was particularly conceived with the aim of providing a bridge of the traditional fields of MEMS and RF MEMS to that of NEMX and its potentialities for enabling opportunities in the emerging field of the IoT and its convergence with mmWave/5G. Having given an in-depth treatment of the fundamental NEMX physics and technology, it is up to the creativity and ingenuity of the scientist or engineer to exploit it to generate new solutions for the IoT and beyond.

Acknowledgements

The idea for this book began to take shape in April, 2016, upon the author receiving a LinkedIn message from Mr. Mark de Jongh who was informing him that he had recently moved from his previous position and started as the Publisher and co-owner at River Publishers. The author had collaborated with Mr. de Jongh in a previous book while he was Senior Publishing Editor at Springer. The author and Mr. de Jongh kept entertaining the idea of the author writing a new book with River Publishers for a couple of years until, finally, with Mr. de Jonhg's encouragement and persistence, the author began the writing in 2018. The author hereby thanks for, and gratefully acknowledges, the opportunity given to him by Mr. de Jongh to write the current book; it has been a pleasure to collaborate with him again. The author also gratefully acknowledges the smooth interactions with Ms. Junko Nakajima during the various production stages. Finally, the author gratefully acknowledges his wife, Violeta, for her understanding along the course of the project.

Héctor J. De Los Santos

List of Figures

List of Tables

List of Abbreviations

5G	fifth generation wireless networks
A	amplitude
ADC	analog-to-digital converter
AFM	Atomic Force Microscope
AlN	Aluminum Nitride
AM	amplitude modulation
ANT	Antenna
BiCMOS	bipolar complementary metal oxide
BLE	Bluetooth Low Energy
BS	base station
CNT	carbon nanotube
CPW	coplanar waveguide
CTE	Coefficient of Thermal Expansion
C-V	capacitance-voltage
D2D	Device-to-Device
DC	direct current
DCO	digitally-controlled oscillator
D/A	digital-to-analog converter
DUT	device under test
EM	electromagnetic
FEA	finite-element analysis
FBAR	Film Bulk Acoustic Wave Resonator
FET	field-effect transistor
FFT	fast Fourier transform
FDTD	Finite-Difference Time-Domain
FF	free-free
FM	frequency modulation
g	gravitational acceleration
G	gain
GHz	giga Hertz
GL	graphene layer
GPa	giga Pascal

GPS	global positioning system
http	Hypertext Transfer Protocol
IEEE	Institute of Electrical and Electronic Engineers
Hz	Hertz
HF	Hydrofluoric acid
IC	integrated circuit
IDT	inter-digitated transducer
IMT	international mobile telecommunications
IoT	Internet of Things
IP	Internet Protocol
ISM	industrial, scientific, and medical applications
kb/s	kilobit per second
km	kilometer
kW	kiloWatt
LAN	Local Area Networks
LDW	Laser direct writing
LLC	logical link control
LO	local oscillator frequency
M2M	machine-to-machine
MAC	media access control
mb/s	megabit per second
MEMS	Microelectromechanical Systems
MIC	microwave integrated circuit
MIMO	Multiple-Input Multiple-Output
MMIC	monolithic microwave integrated circuit
MPa	mega Pascal
mV	milliVolt
N	Newton
NAN	neighborhood area network
NEMX	Nanoelectromechanical Quantum Circuits and Systems
nF	nanoFarad
nm	nanometer
ns	nanosecond
NEMS	Nanoelectromechanical Systems
mmWave	millimeter wave
OFET	Organic Field-Effect Transistor
OLED	Organic Light Emitting Diode
OPD	Organic Photo Diode Organic Light
OSC	organic solar cell

OSI	Open Systems Interconnection
PAN	personal area network
PC	Personal Computer
PE	printed electronics
PECVD	plasma-enhanced chemical vapor deposition
PEC	perfect electrical conductor
PFD	power flux density
PLL	phase-locked loop
PDMS	poly-dimethylsiloxane
PZT	Lead Zirconate Titanate
PVDF	polyvinyledene fluoride
Q	quality factor
RC	resistor-capacitor
RF	Radio-Frequency
RFID	radio-frequency identification device
RL	resistor-inductor
RLC	resistor-capacitor-inductor
RMS	room-mean-square
RN	random number
RX	receiver
Scope	oscilloscope
SEM	scanning electron micrograph
STR	simultaneous transmission and reception
SWCNT	single-wall carbon nanotube
TCP	Transmission Control Protocol
TEM	transverse electromagnetic
TFT	Thin-Film Transistor
TMax	maximum deposition temperature
TPa	Tera Pascal
TR	tuning range
TV	television
TX	transmitter
UV	ultra-violet
VCO	voltage controlled oscillator
VGB	gate-beam actuation voltage
VHF	very high frequency
VNI	visual network index
WAN	wide area network

1

The Internet of Things

1.1 Origins

What is the Internet of Things (IoT)? The term was proposed in 1999 by Kevin Ashton, a British Technologist, when he was at the Massachusetts Institute of Technology (MIT), in the context of exploring the potential of networked radio-frequency identification devices (RFIDs) and emerging sensing technologies [1]. He defined the IoT as follows: "The Internet of Things connects devices such as everyday consumer objects and industrial equipment onto the network, enabling information gathering and management of these devices via software to increase efficiency, enable new services, or achieve other health, safety, or environmental benefits."

A more recent definition, by David Evans at Cisco Systems [2], states that: "The IoT refers to a network of smart devices communicating and exchanging data with other machines, objects, devices and environment around the globe."

In terms of its chronological development, the IoT has been defined as having been born between the years 2008 and 2009, when the number of connected devices became greater than the number of people in the Earth (Table 1.1).

The IoT may be viewed as a network of networks, where we have individual networks connected together within a system endowed with security, analytics, and management (Figure 1.1).

As suggested by Figure 1.1, the IoT has such an enormous breadth that it can be difficult to understand or comprehend. To facilitate its comprehension, the IoT may be broken down into five key vertical categories of adoption, namely connected wearable devices, connected cars, connected homes, connected cities, and the industrial Internet (Figure 1.2) [1].

Table 1.1 Perspective of IoT evolution

World Population	6.3 Billion		6.8 Billion	7.2 Billion	7.8 Billion
Connected devices	800 Million		12.6 Billion	25 Billion	50 Billion
Connected devices per person	0.08	More connected devices than people	1.84	3.47	6.58
Year	2003		2010	2015	2020

Source: Ref. [2].

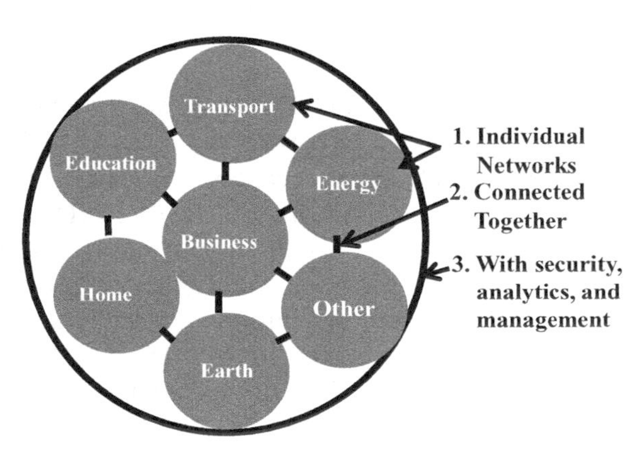

Figure 1.1 The IoT as a network of networks.

Source: Ref. [1].

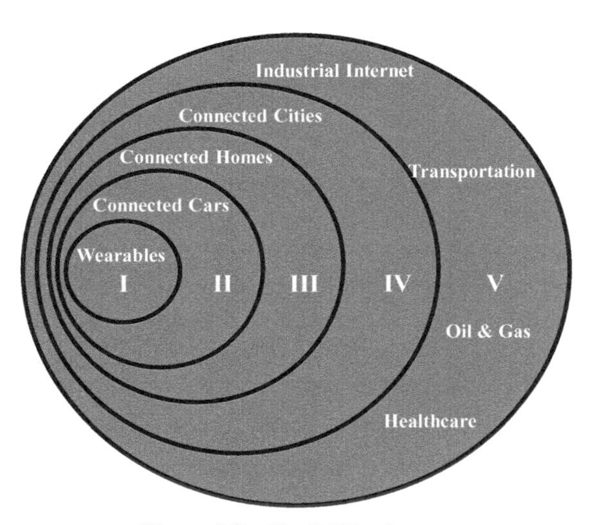

Figure 1.2 The IoT landscape.

Source: Ref. [1].

1.2 IoT Motivation/Impact

The projection of the number of connected devices to which the IoT application categories will give rise is shown in Figure 1.3 [1].

Detailed consideration of the IoT makes it clear that its realization may well rest upon the following three fundamental pillars: (i) sensing, (ii) wireless communications, and (iii) security.

Sensing encompasses every imaginable device or system that can capture information from the environment or otherwise. This information, in turn, being of a diverse and distributed (ubiquitous) nature, must be gathered and disseminated throughout a dispersed network of intelligent nodes, which motivates the utilization of wireless communications, the most efficient means to carry out this task. Finally, since the information being distributed to the end users will include extremely valuable information, that is, not only that of a personal nature but, in particular, that of a business nature (e.g., company finances, bank accounts, and credit cards), and that pertaining to the control of autonomous and remotely controlled vehicles, in the context of the Internet, schemes to enable secure, hacking-free wireless communications techniques, must be employed. Traditionally, secure communications has relied upon the generation of random numbers (RNs), which are employed for effecting information modulation or cryptographic encoding in such a way that the information in question can only be demodulated or decoded with a replica of such RNs [3]. In this context, since the RNs are generated by deterministic computer algorithms that eventually repeat the RN sequence, a risk exists that such RNs may be deciphered and used to extract the information transmitted. To overcome this risk, a number of approaches to generate RNs via physical processes in devices that, being derived from naturally occurring physical effects and processes, do not repeat themselves

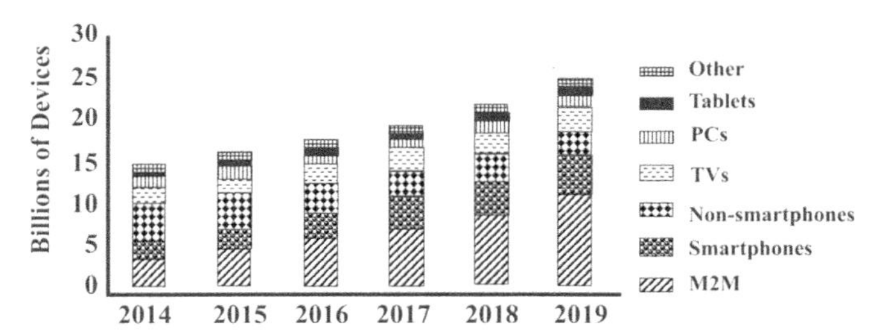

Figure 1.3 Global projected growth of connected devices by type.

Source: Ref. [1].

and thus are purely (ideally) random and are of current research interest for IoT applications.

A wide variety of sensors will make the IoT possible (Figure 1.4). These are typified by microphones, gyroscopes, accelerometers, pressure sensors, magnetic sensors, and so on, which may be attached to mobile devices [4]. It may easily be noticed that a common denominator that underpins the majority of these sensors is microelectromechanical/nanoelectromechanical systems (MEMS/NEMS) technology [5, 6].

The main sensor drivers for consumer electronics, for instance, include motion detection, pedestrian navigation, position detection, and user interface (Figure 1.5).

Taking the potential of the IoT paradigm to the extreme, Hewlett-Packard and others have proposed "the central nervous system of the Earth's" vision, which entertains the fusion of man, machine, virtual, and physical

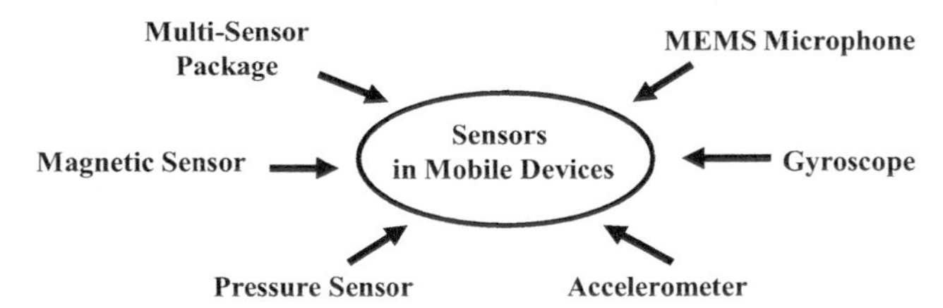

Figure 1.4 Sensors in mobile devices for the Internet of Things.

Source: Ref. [4].

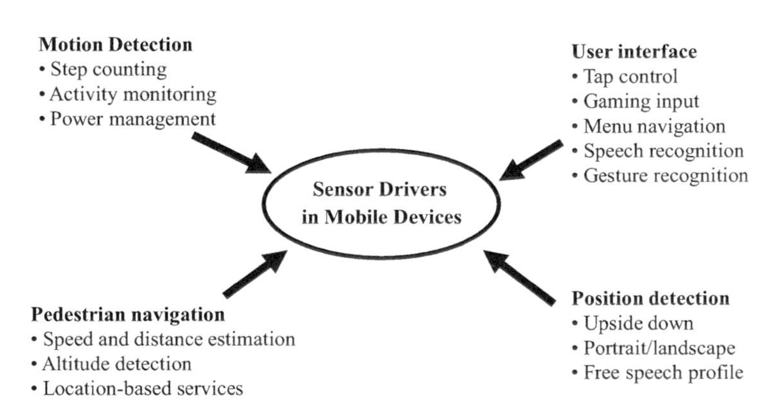

Figure 1.5 Sensors drivers in mobile devices for the Internet of Things.

Source: Ref. [4].

systems to revolutionize human interaction with the Earth in as profound a fashion as the Internet has revolutionized personal and business interactions (Figure 1.6) [7]. Upon coming to fruition, this scenario would entail approximately one trillion nanoscale devices, that is, the equivalent number of sensors and actuators of 1000 Internets.

In the final analysis, one of the key drivers of the Internet of Things is its impact on the world economy. In that regard, it is estimated that the IoT will give rise to a market larger than that of the personal computer (PC), tablet,

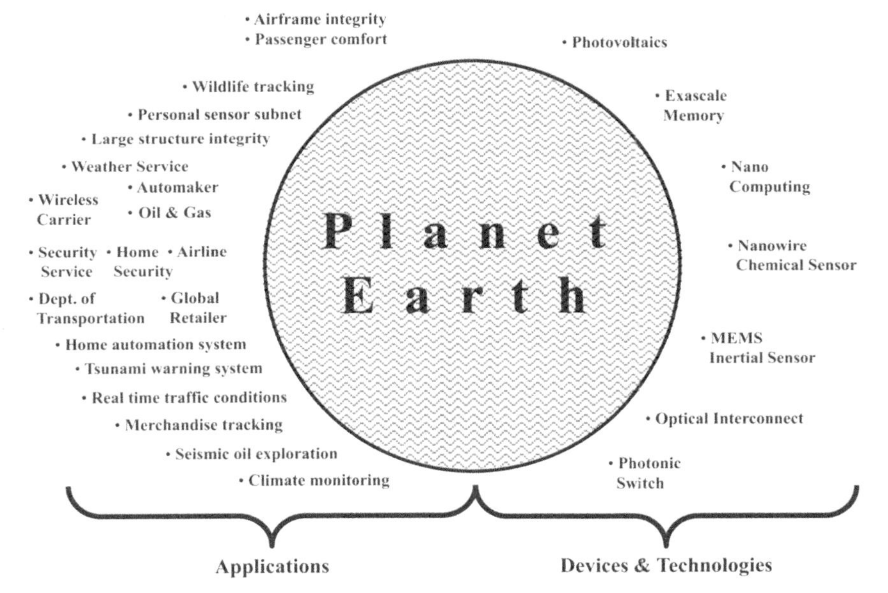

Figure 1.6 The central nervous system of the Earth vision.

Source: Ref. [7].

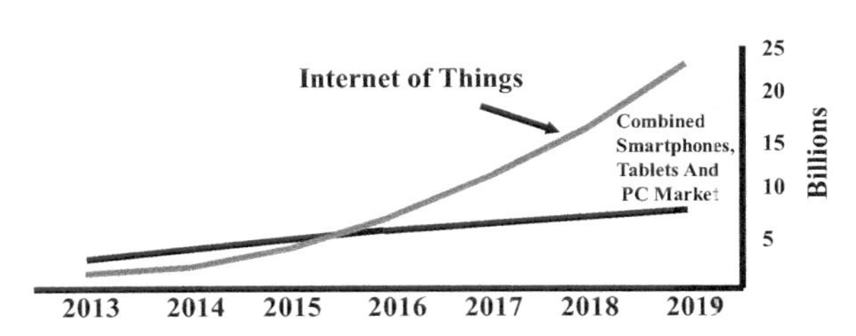

Figure 1.7 Projected IoT device growth versus PC, smart phone, and tablet growth.

Source: Ref. [8].

and smart phone markets combined of the order of greater than $20B/year (Figure 1.7) [8]. Not only, therefore, does the development of the IoT portend orders of magnitude improvements in our way of life but it also heralds the potential for great technological and economic development and prosperity.

1.3 Summary

In this chapter, we have presented the origins, motivation, and potential impact of the Internet of Things. The IoT crystallizes the large-scale integration of advances in sensors, wireless communications, and networking ushered by the growth in Internet connectivity over the last twenty plus years. At the core level, this growth will be fueled by the proliferation of a virtually innumerable set of intelligent sensor nodes that can communicate via secure, wireless links over a worldwide network of networks. In the rest of this book, we will endeavor to open the reader's understanding of the key technological enablers of the IoT, with the goal of equipping him or her with the fundamental knowledge base to, if desired, become a participant and contributor to the grand IoT challenge.

2

Microelectromechanical and Nanoelectromechanical Systems

2.1 MEMS/NEMS Origins

It can be said that the field of microelectromechanical/nanoelectromechanical systems (MEMS/NEMS) may be traced back to Richard Feynman when, in 1959, he made the observation: "There is plenty of room at the bottom." [9]. Feynman reached this conclusion upon conducting a special type of search, namely a search for a boundless field. He noticed that fields like endeavoring to attain low temperatures or attaining high pressures had virtually no end in sight. That is, one could never say or claim that one had reached the lowest temperature or the highest pressure. Feynman's inquiry about a new boundless field led him to that of *miniaturization*. Indeed, endeavoring to make everything small was a research field that had virtually no end in sight [10].

But why had so little been done on miniaturization? He wondered. Were there innate limitations imposed by the laws of physics? No, Feynman concluded. There is nothing in the laws of physics that precludes miniaturization. Rather, the limitations are rooted in technology, that is, it is one's ability to make small things that sets a limit on miniaturization. Having acknowledged that miniaturization was limited by technology and that this limitation would erode in time, Feynman went on to consider what if the technological problem/limitation was nonexistent? How would miniaturization impact the three particular areas of application, namely information storage, computers, and, more pertinent to this book, machinery? His assessment of the impact of miniaturization led him to predict that great gains in information storage and computer systems would be achieved by straightforward downscaling, since these functions did not depend on size, but that, in turn, the scaling down of machines would require new paradigms because machine properties, in particular, the force systems involved, were

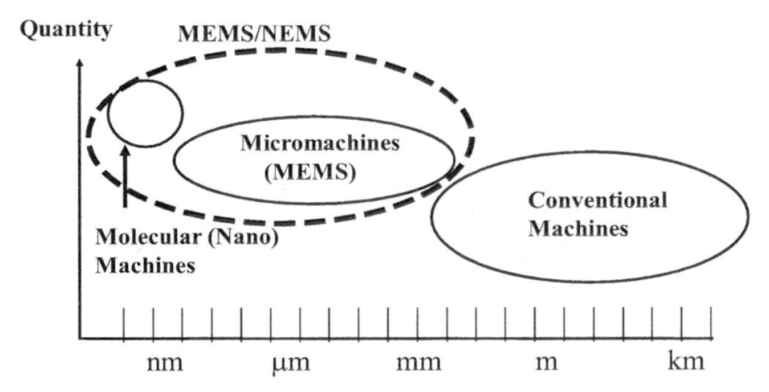

Figure 2.1 Domain of machinery sizes.

Source: Ref. [11].

constrained by size. As a result, we would be confronted by new regimes of material behavior, strange to our everyday experience [6], and grappling with this new domain, he concluded, would open up a whole new field of scientific endeavor. Overall, he suggested, miniaturization would fuel radical paradigm shifts in all areas of science and technology, and these, in turn, would elicit a virtually unlimited amount of applications.

But what did Feynman mean by small? An examination of the machines in operation back in 1959 reveals that they belonged to two-dimensional regimes typified, on the one hand, by machines comprising sizes from millimeters to kilometers and, on the other hand, by molecular machines, biological machines, which could be synthesized by chemists, occupying the sub-nanometer scale. In-between, there was a big empty space! This space was the regime to be tackled by MEMS/NEMS (Figure 2.1). Notice that this dimensional regime was concomitant with that utilized with technology for making electronic devices [12].

2.2 MEMS/NEMS Impetus/Motivation

The field of miniaturization lay virtually dormant until our ability to make small things improved, which began to take place in the 1960s with the advent of integrated circuit (IC) fabrication technology. In particular, since circuits could be scaled down in size and still perform their function, a race ensued to develop ways to print more and more circuits on a substrate (wafer). This downscaling thrust was, on the economic side, beneficial because the greater the number of circuits that could be formed in a given wafer area, the greater were the profits. This behavior, which characterized the outstanding economic success and progress of the semiconductor industry, came to be captured by the famous Moore's law [13]. In fact, if one uses the number of devices as an index of the extent to which integration or miniaturization has progressed, one finds that the number of devices on a wafer has increased by more than seven orders of magnitude, namely from <10 in the 1960s to >1 billion today [14].

The economic success in the IC industry led people to wonder: Would the application of IC fabrication concepts in other fields, like mechanics, optics and fluidics, result in enhanced performance and reduced cost? The answer was maybe, because whereas an IC extends in two dimensions (2D), a mechanical structure is 3D in nature. Therefore, techniques had to be developed for microstructure generation, which would also allow the third dimension of a structure to be shaped. Beginning with IC fabrication (2D) technology, a summary of the fundamentals of fabricating 3D structures is given in Appendix A.

MEMS/NEMS may be utilized to exploit a multitude of physical phenomena (Figure 2.2). Indeed, the universe of possible implementations is vast and is only limited by our imagination.

Possible Physical Mechanisms to be Exploited

- Acoustical
- Electrical
- Optical
- Mechanical
- Magnetic
- Fluidic
- Quantum Effects
- Mixed Domain

Some Areas of Endeavor Under R&D

- Nanoelectronics
- Nanomechanics
- Nanoengineering
- Nanobiotechnology
- Nanomedicine
- RF MEMS
- IoT

Figure 2.2 Physical phenomena available for exploitation in MEMS/NEMS and potential areas of application.

2.3 Summary

Indeed, the potentialities of the MEMS/NEMS field, as suggested by Feynman in 1959, are virtually unlimited. These potentialities, due to the astounding progress in miniaturization over the last 40 plus years, captured by the Moore's law and exemplified by the ubiquity of IC technology, are ushering a new technological revolution. This revolution is enabled by the convergence, on the one hand, of the ability of MEMS/NEMS and IC fabrication technologies to realize miniaturized systems capable of sensing, processing, and knowledge wireless distribution through the Internet and, on the other hand, the emergence of new applications demanding virtual *omniscience*, namely the *Internet of Things* (IoT). In the next chapter, we develop a fundamental understanding of some key physical principles upon which MEMS/NEMS devices for the IoT are predicated.

3

Understanding MEMS/NEMS Device Physics

3.1 Actuation

It is well known that Physics is the natural science that deals with the study of matter and its motion and behavior through space as a manifestation of the energy it possesses and the forces it experiences [15]. Microelectromechanical system (MEMS)/nanoelectromechanical system (NEMS) device physics deals with the behavior of miniaturized mechanical and electronic/plasmonic [16] devices engineered to move or transmit their motion as a result of an applied voltage or current. The resulting device motion caused is referred to as the *actuation* of the device, and the devices transmitting the force are referred to as *actuators* [5].

In the realm of MEMS/NEMS dimensional regimes, a wide variety of forces may be exploited to cause actuation, namely from those derived from electrostatic, piezoelectric, magnetic, or thermoelectromechanical means to those derived from the quantum vacuum (Casimir forces) or optical light beams (radiation pressure). In this subsection, we present the fundamentals of the most important MEMS/NEMS actuation mechanisms of relevance to IoT.

3.1.1 Electrostatic Actuation

3.1.1.1 Parallel-plate capacitor

Electrostatic forces derive from the variation of the energy stored in a capacitor containing positively and negatively charged metallic plates, as a result of the interaction of the plates' charges that arises from the application of a voltage across it [5]. How this actuation comes about may be illustrated by considering the capacitor in Figure 3.1. In the usual case, when both the top and bottom plates are rigid (unmovable and unbendable), if the plate area is much greater than the separation between them, so that the fringing

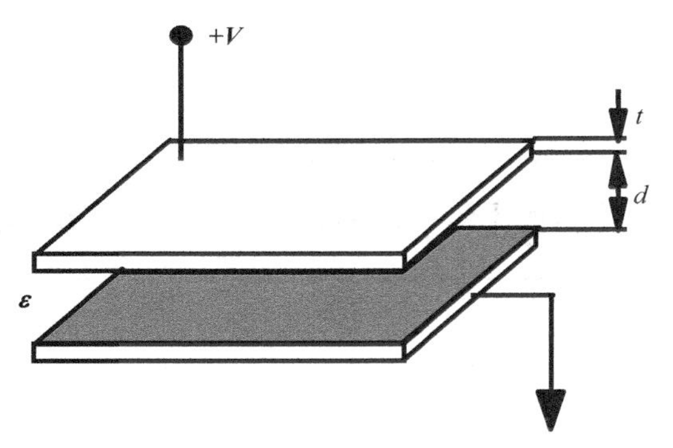

Figure 3.1 The parallel-plate capacitor embodies the fundamental elements of the electrostatic actuator. ε represents the permittivity between the plates, A represents the area of the plates, d represents the separation between the plates, and t represents the thickness of the plates.

capacitance may be ignored, the capacitance is given by [17],

$$C = \frac{\varepsilon A}{d} \tag{3.1}$$

where ε represents the permittivity between the plates, A is the area of the plates, and d is the separation between the plates.

Application of a voltage V to the capacitor of value C gives rise to the storage of an electrostatic potential energy U within the inter-plate volume, expressed by [17],

$$U = \frac{1}{2}CV^2 = \frac{\varepsilon AV^2}{2d} \tag{3.2}$$

This potential energy represents the electrostatic Coulomb force of *attraction* of plates situated at a fixed distance d, carrying respective positive (q_P) and negative (q_N) charges, given by [17],

$$F = \frac{1}{4\pi\varepsilon}\frac{q_P q_N}{d^2} \tag{3.3}$$

Electrostatic force-induced actuation/motion in a parallel-plate capacitor manifests itself when, under the application of a varying voltage, one lifts the plate rigidity limitation. In this case, a voltage increase causes a reduction in the separation d (and, thus, an increase in the capacitance C) and, consequently, an increase in the stored potential energy U. This sequence of events establishes the conditions for a positive feedback system in the sense that one thing, namely reduction in d, causes the increase in another, namely the potential energy, which by increasing the Coulomb force reinforces the first. Because under this circumstance the potential energy is variable (with varying applied voltage), the instantaneous attractive force capturing the positive feedback action is given by [17],

$$F = -\nabla U \tag{3.4}$$

or substituting Equation (3.2) into Equation (3.4),

$$F = \frac{\varepsilon A V^2}{2d^2} \tag{3.5}$$

Now, assume that the bottom plate of the capacitor is attached/bonded onto the substrate and thus immovable and unbendable. Then, electrostatic actuation in the parallel-plate capacitor manifests itself when, under the application of a voltage, the top plate is allowed to move, to bend, or both. The situation in which the top plate is movable, but rigid (unbendable) may be represented as in Figure 3.2.

Here, the rigid (assumed unbendable) top plate moves a uniform distance x, which is a function of the applied voltage, in such a way that the instantaneous equilibrium position x is given by the solution to the equation, $|F_S| = |F_E|$, expressed in Equation (3.6) and shown in Figure 3.3 [18].

$$F_S = kx = F_E = \frac{\varepsilon A V^2}{2(d-x)^2} \tag{3.6}$$

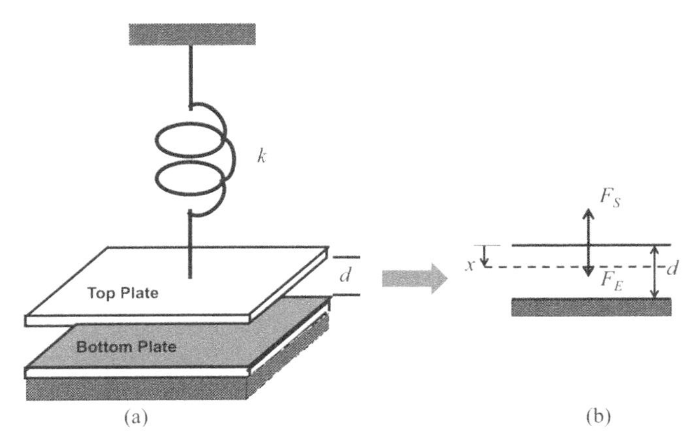

Figure 3.2 (a) Parallel-plate capacitor with bottom plate fixed onto substrate and unbendable top plate supported by a spring of spring constant k. (b) Forces on parallel-plate capacitor of (a). F_S represents the spring force and F_E represents the electrostatic force upon the top plate reaching a position of equilibrium, x, due to an applied voltage.

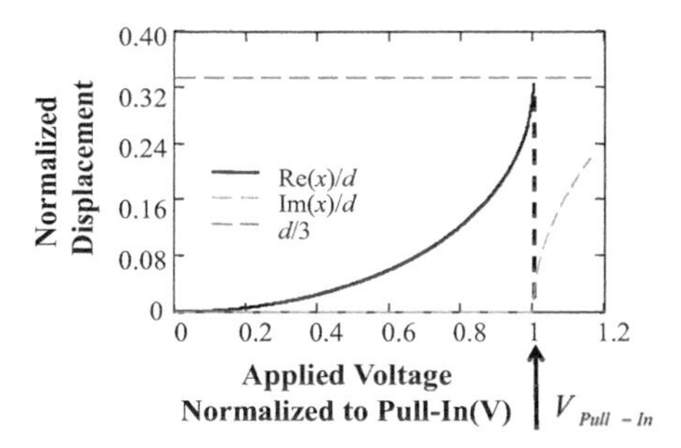

Figure 3.3 Normalized displacement of rigid (unbendable) capacitor top plate versus normalized applied voltage.

Source: Ref. [18].

The solution to Equation (3.2) is shown to be real until the displacement x reaches $d/3$, at which point it becomes imaginary. The applied voltage at which $x = d/3$ is referred to as the *pull-in* voltage [5]. It represents the voltage at which the electrostatic attraction Coulomb force can no longer be countered or equilibrated by the spring force and the top plate crashes down on the bottom plate.

An analytical expression for the pull-in voltage may be obtained by solving for the voltage for which the difference between the spring and electrostatic forces,

$$\left(kx - \frac{\varepsilon A V^2}{2(d-x)^2}\right) = 0 \tag{3.7}$$

is a minimum. From equating to zero, the derivative of the voltage with respect to the displacement, Equation (3.8), and solving for V at $x = d/3$, one obtains,

$$\frac{dV}{dx} = \frac{d}{dx}\left[\sqrt{\frac{2kx(d-x)^2}{\varepsilon A}}\right] = 0 \rightarrow x = \frac{d}{3} \tag{3.8}$$

giving,

$$V_{Pi} = \sqrt{\frac{8kd^3}{27\varepsilon A}} \tag{3.9}$$

3.1.1.2 Electrostatically actuated cantilever beam

If one assumes that, instead of being supported by a spring, the top plate is anchored on one side and is not rigid, but bendable/flexible, that is, that

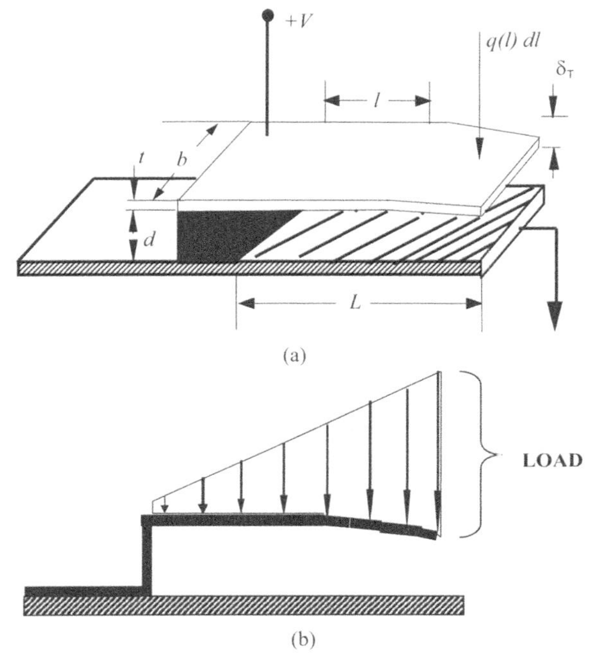

Figure 3.4 (a) Parallel-plate capacitor in which the top plate is flexible/bendable and is anchored on one side. (b) Load distribution causing the displacement of the top plate to increase from zero at the anchor to a maximum δ_T at its free tip.

Source: Ref. [18].

we have a cantilever-type structure, then we have the scenario depicted in Figure 3.4 [18]. Here, since the attractive electrostatic force cannot cause any motion/displacement of the top plate at the anchor, it adopts a *deformed shape* which is embodied by the position-dependent displacement $\delta(l)$. This displacement grows from $\delta(l = 0) = 0$ at the anchor to a maximum of $\delta_T(l = L)$ at the free tip. The force experienced by the top plate at a distance l from the anchor, in particular, is given by,

$$q(l) = \frac{\varepsilon_0 A V^2}{2(d - \delta(l))^2} \tag{3.10}$$

where $\delta(l)$ embodies the deflection curve exhibited by the deformed beam, whose mechanical characteristics are a function of the beam's structural and material properties captured by its Young's modulus, E, and its moment of inertia, I(Figure 3.4b). The, now classical, analysis of this situation by

Petersen [19] posits, on the one hand, that the concentrated load on a cantilever beam at a distance l from the anchor gives rise to a beam tip displacement given by,

$$\delta_T(l) = \left[\frac{l^2}{6EI}\right](3L - 1)bq(l)dl \tag{3.11}$$

where b is the beam width. On the other hand, it is assumed that the position-dependent displacement is parabolic and given by,

$$\delta(l) = \left(\frac{l}{L}\right)^2 \delta_T \tag{3.12}$$

The forces distributed along the length of the beam and the resulting deflection at the tip are then found by integrating Equation (3.11) from $l = 0$ to $l = L$, that is,

$$\delta_T = b\int_0^L \frac{3L - l}{6EI} l^2 q(l)dl \tag{3.13}$$

The integral in Equation (3.13) was solved to give a normalized load, defined in the below equation,

$$l = \frac{\varepsilon b L^4 V^2}{2EId^3} \tag{3.14}$$

The normalized deflection at the tip of the cantilever beam is expressed by,

$$l = 4\Delta^2 \left[\left(\frac{2}{3(1-\Delta)}\right) - \frac{\tanh^{-1}\sqrt{\Delta}}{\sqrt{\Delta}} - \frac{\ln(1-\Delta)}{3\Delta}\right]^{-1} \tag{3.15}$$

with,

$$\Delta = \frac{\delta_T}{d} \tag{3.16}$$

In analogy with the parallel-plate capacitor case, it is found from an examination of a plot of Equation (3.15) that there exists a threshold bias at which the beam, no longer controlled by the applied bias, crashes down on the substrate. This bias point is again reached after the beam deflects about a distance $\delta_{\mathrm{T}} = d/3$. This collapse was explained by Petersen [19] as being a result of increasing concentration of the electrostatic forces at the tip, so that at a particular voltage, this concentrated load causes the beam position to become unstable, and it undergoes a spontaneous deflection, the remaining distance.

When the operation of the cantilever beam includes it being driven into the instability regime, the phenomenon of *hysteresis* is observed. In particular, once the beam is fully deflected, the subsequent reduction of the applied voltage has no effect on its state of deflection until it becomes lower than the threshold. Concise expressions to capture hysteresis were derived by Zavracky et al. [20], including expressions for the beam closing (threshold) and opening voltages in terms of the effective beam spring constant, $K = bt^3E/4l^3$, the beam area A, and the initial and effective closed beam-to-electrode distances, d and dc, respectively. Accordingly, the closing and opening voltages are given by Equations (3.17) and (3.18), respectively.

$$V_{th\text{-}close} = \frac{2}{3}d\sqrt{\frac{2Kd}{3\varepsilon_0 A}} \tag{3.17}$$

$$V_{th\text{-}open} = (d - d_c)\sqrt{\frac{2Kd_c}{3\varepsilon_0 A}} \tag{3.18}$$

3.1.1.3 Interdigitated (comb-drive) capacitor

The *comb-drive* capacitor is another form of electrostatic actuator. In this architecture, the application of a voltage controls the degree of displacement, x, which is the distance a moveable frame (the "rotor") containing a set of "teeth" engages into a fixed/static electrode frame (the "stator") flanking each tooth at a lateral distance g(Figure 3.5).

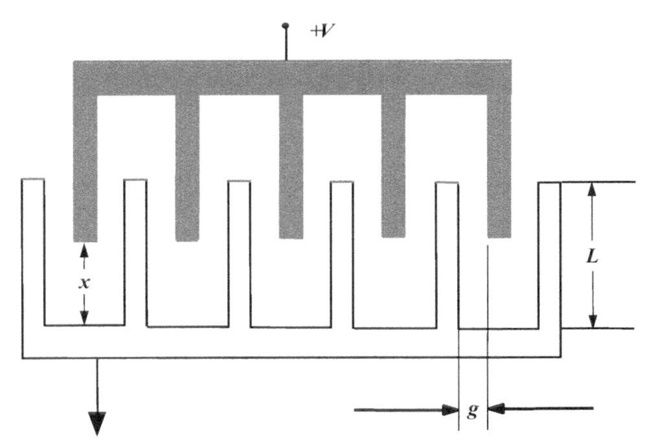

Figure 3.5 Interdigitated comb-drive actuator. The thickness into the plane is "t".
Source: Ref. [18].

In this case, if the fingers have a thickness t and a length L, then the single-finger capacitance area is given by,

$$A = t(L - x) \tag{3.19}$$

from where the single-finger capacitance is given by,

$$C_{single} = \frac{\varepsilon A}{g} = \frac{\varepsilon t(L - x)}{g} \tag{3.20}$$

Due to the fact that each tooth has two sides, it follows that each tooth has two capacitors. Thus, for an n teeth rotor, we have $2n$ capacitors and the total capacitance is,

$$C_{single} = 2n\frac{\varepsilon t(L - x)}{g} \tag{3.21}$$

Following the procedure in Equation (3.1) through Equation (3.4), but using capacitance Equation (3.21), we obtain the comb-drive actuation force–displacement relationship in Equation (3.22),

$$F = n\varepsilon\frac{t}{g}V^2 \tag{3.22}$$

Examination of Equations (3.5) and (3.22) reveals that while for a parallel-plate capacitor the force varies as $1/x^2$, for the comb-drive device it is constant independent of x [18]. One aspect to be aware of in designing comb-drive actuators is that, due to fringing fields, forces out of the plane arise which can result in levitation of the actuator away from the substrate [18]. Furthermore, if the lateral stiffness is insufficient, there may occur a lateral instability, depending on how the actuator is supported. This manifests as a tendency of the rotor to be attracted sideways and its teeth to stick to those of the stator.

3.1.2 Piezoelectric Actuation

In the piezoelectric mechanism of actuation, use is made of the deformation of structures as a result of the motion of *internal* charges elicited by an applied electric field. Conversely, an applied stress on a piezoelectric structure elicits an electric field in it as a result of the forced motion of the internal charges. Because of the anisotropic properties of piezoelectric crystals, there is coupling between electric fields and strains in different directions. A manifestation of this behavior, for instance, is seen when a piezoelectric crystal with an imposed z-directed electric field exhibits a strain in the x-direction. This phenomenon is expressed by the following set of constitutive equations [18]:

$$\begin{aligned} S_x &= s_{xx}^E T_x + d_{zx} E_z \\ D_z &= d_{zx} T_x + \varepsilon_{zz}^T E_z \end{aligned} \tag{3.23}$$

These equations are interpreted as follows. Firstly, the strain in the x-direction, s_x, results from the sum of two strains, namely that elicited by an x-directed stress, T_x, in the absence of an electric field, E_z, and that elicited by an electric field, E_z, in the absence of an x-directed stress, T_x. Secondly, the z-directed electric displacement, L_z, is elicited by an x-directed stress, T_x, in the absence of an electric field, E_z, and that elicited by an electric field, E_z, in the absence of a stress, T_x. The stress, T_x, and the electric field, E_z, produce the strain, s_x, and the electric displacement, L_z, via the following constitutive parameters: The elastic compliance in the x-direction due to an x-directed stress, in the absence of an electric field, s_{xx}^E, the piezoelectric coupling coefficient, relating a z-directed electric field, E_z, to an x-directed strain, d_{zx}, and the z-directed permittivity at constant stress due to a z-directed electric field, ε_{zz}^T.

3.1.2.1 Piezoelectric cantilever probe

The prototypical MEMS actuator, embodying piezoelectric actuation, is the piezoelectric cantilever beam (Figure 3.6). In this prototype, a piezoelectric material, which is realized by a composite-layer structure, is deposited on a beam. In particular, the piezoelectric material, that is, a capacitor, is sandwiched between two electrodes. To elicit actuation, a voltage set across the capacitor produces an electric field in the z-direction which, in turn, produces a strain/elongation of the piezoelectric layer in the x-direction. Because the beam itself is not piezoelectric, its size doesn't change and the result of the piezoelectric material elongation is bending, in other words, its tip displaces in the z-direction. How large a displacement is created depends on the lateral width of the beam, with a value of several microns per 100 μm of beam width being typical, together with a change in beam thickness of 0.1 nm per Volt [21]. Typical piezoelectric materials employed in MEMS devices include zinc oxide (ZnO), aluminum nitride (AlN), lead zirconate titanate ($PbZr_xTi_{1-x}O_3$ – or more commonly, PZT), and polyvinyledene fluoride (PVDF).

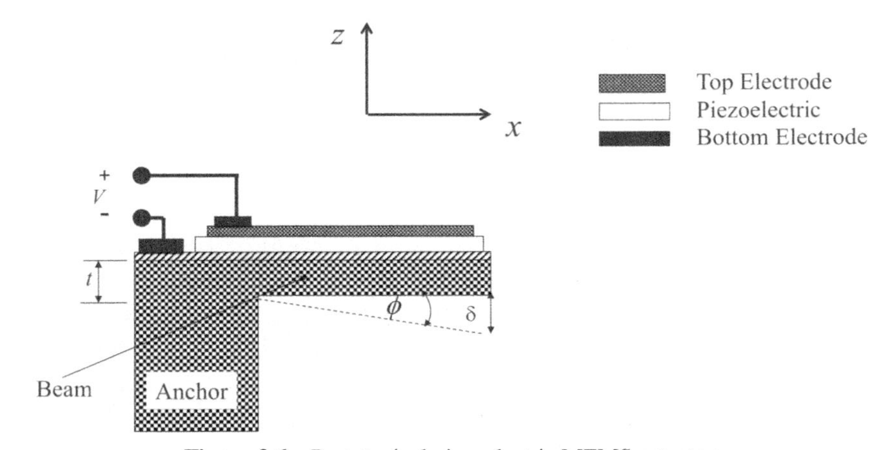

Figure 3.6 Prototypical piezoelectric MEMS actuator.

Source: Ref. [21].

Various analyses of the motion of an actuated piezoelectric multi-morph cantilever beam have been undertaken, from detailed, for example, Weinberg [22], to simplified, for example, DeVoe [23] and Bashir [24]. Taking a thin piezoelectric layer on a thick beam, the beam tip displacement and the angle of rotation are given as follows [24]:

$$\delta = 3d_{zx}\frac{L^2 E_p}{t^2 E}V \tag{3.24}$$

$$\theta = 6d_{31}\frac{LE_p}{t^2 E}V \tag{3.25}$$

where L is the beam length, t indicates its thickness, E_P indicates the piezoelectric layer Young's modulus, and E indicates the beam Young's modulus, and V indicates an applied voltage, using these equations one can obtain tip deflection and rotation δ and θ, respectively. For a beam of cross-sectional area A, mass density ρ, length L, and moment of inertia I, the corresponding fundamental mechanical bending resonance frequency is given by,

$$\omega_0 = 3.5160\left(\frac{EI}{\rho L^4 A}\right)^{1/2} \tag{3.26}$$

3.1.3 Casimir Actuation

So far, the actuation mechanisms discussed have relied on the application of a voltage. Other actuation mechanisms such as magnetic and thermoelectromechanical [18] may also be exploited. We have limited the discussion to electrostatic and piezoelectric actuation, however, because they are compatible, tend to exhibit low power consumption, and are easily implemented in the context of integrated circuit (IC) technology. One actuation mechanism that is independent of any applied bias will now be introduced.

When the proximity between material objects reaches the few microns and below, a regime is entered in which forces that are quantum mechanical in nature [25–29], namely van der Waals and Casimir forces, become operative. These forces supplement, for instance, the electrostatic force in countering Hooke's spring force to determine the beam actuation behavior. They may also be responsible for stiction [30], that is, causing close interaction of elements (Figure 3.7), to adhere together and, thus, may profoundly change actuation dynamics [31, 60]. Next, we present several approaches to derive the Casimir force that provide insight into the nature of this force and intuition into how to deal with it.

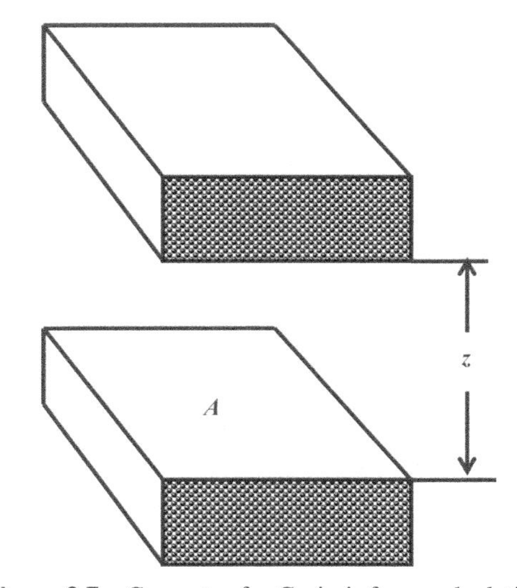

Figure 3.7 Geometry for Casimir force calculation.

3.1.3.1 Casimir's own force calculation

The Casimir force has its origin in the polarization of adjacent material bodies that are separated by distances of less than a few microns and is the result of *quantum-mechanical fluctuations* in the electromagnetic field permeating the free space between them [25–29]. It may also be traced to the vacuum fluctuations of a classical real electromagnetic field [32]. In particular, when the material bodies in question are parallel conducting plates, separated by free space, the Casimir force is attractive [25]. However, in general whether the force is attractive or repulsive [33–35] is a function of both the boundary conditions, in particular, of the specific geometry sampled by the field and the relationship among the material properties of the plates and the intervening space.

The surprising aspect of the Casimir force is that it is a macroscopic observable of the purely quantum-mechanical prediction of *zero-point vacuum fluctuations* [6]. Thus, even when the average electromagnetic field is zero, its average energy shows fluctuations with *small* but non-zero value. The fundamental calculation of the Casimir force entails computing the energy between the plates with perfectly smooth surfaces and obtaining its gradient. Since the zero-point vacuum energy, $E_{Field} = \frac{1}{2}\hbar \sum_n \omega_n$, diverges, many techniques have been developed to accomplish this calculation [36, 37]. The essence of many of these calculations, however, is to compute the physical energy as a difference in energy corresponding to two different geometries, for example, the parallel plates at a distance "*a*" apart and the parallel plates at a distance "*b*," where the limit as *b* tends to infinity is taken. For flat surfaces, the infinite part of the energy cancels when the energy difference of the two configurations is taken.

Casimir [25, 50] began by assuming a cubic box of volume L^3 bounded by perfectly conducting walls and a perfectly conducting square plate placed inside it, parallel to xy face. Then, he calculated the zero-point energy when the plate was both a short (I) and a large (II) distance $L/2$ from the xy plane. He noticed that in both cases, expressions $\frac{1}{2}\sum \hbar\omega$, where the summation extends over all resonance frequencies of the cavities, are divergent and devoid of physical meaning, but their difference, $\delta E = \left(\frac{1}{2}\sum \hbar\omega\right)_I - \left(\frac{1}{2}\sum \hbar\omega\right)_{II}$, did (*and does*) have a definite value that can be interpreted as the interaction between the plate and the face. In his calculation, the zero-point energies were expressed as [25],

$$\frac{1}{2}\sum \hbar\omega = \hbar c \frac{L^2}{\pi^2} \int_0^\infty \int_0^\infty \left[\frac{1}{2}\sqrt{k_x^2 + k_y^2} + \sum_{n=1}^{\infty} \sqrt{n^2 \frac{\pi^2}{z^2} k_x^2 + k_y^2} \right] dk_x dk_y \tag{3.27}$$

where, for plates of area $L \times L$ and separation z, $k_x = n_x\pi/L, k_y = n_y\pi/L$, and $k_z = n_z\pi/z$. In particular, to every k_x, k_y, and k_z, there correspond *two* standing waves, unless n_i is zero, in which case there is only one. For large L, this has no consequence in k_x and k_y, because they behave as continuous. By going to polar coordinates in the transverse (xy plane, represented by $\aleph$), Casimir expressed Equation (3.27) in one single term as,

$$\frac{1}{2}\sum \hbar\omega = \hbar c\frac{L^2}{\pi^2}\cdot\frac{\pi}{2}\sum_{(0)1}^{\infty}\int_0^{\infty}\left(\sqrt{n^2\frac{\pi^2}{z^2}+\aleph^2}\right)\aleph d\aleph \tag{3.28}$$

where the notation (0)1 is meant to indicate that, in the term $n = 0$, a factor of 1/2 must be introduced.

This expression is general, that is, it applies for any inter-plate separation z. For large separations, z, the sum in Equation (3.28) may be converted to an integral, and the expression for the change in zero-point energy becomes,

$$\begin{aligned}\delta E &= \hbar c\frac{L^2}{\pi^2}\cdot\frac{\pi}{2}\left\{\sum_{(0)1}^{\infty}\int_0^{\infty}\left(\sqrt{n^2\frac{\pi^2}{z^2}+\aleph^2}\right)\aleph d\aleph\right.\\ &\left.-\int_0^{\infty}\int_0^{\infty}\sqrt{k_z^2+\aleph^2}\aleph d\aleph\left(\frac{z}{\pi}dk_z\right)\right\}\end{aligned} \tag{3.29}$$

To obtain the final result, Casimir multiplied the integrand in Equation (29) by a function $f(k/k_m)$, that is, unity for $k \ll k_m$, but tends to zero for $k \gg k_m \to \infty$, where k_m is defined as $f(1) = 1/2$. This was the introduction of the famous wavelength cutoff, with the physical meaning that: For short wavelengths, the plates are not an obstacle, and consequently, the zero-point energy is not influenced by their position. To continue, Casimir made the change of variable $u = z^2\aleph^2/\pi^2$ in Equation (3.29), which allowed him to express it as,

$$\begin{aligned}\delta E &= L^2\hbar c\frac{\pi^2}{4z^3}\left\{\sum_{(0)1}^{\infty}\int_0^{\infty}\left(\sqrt{n^2+u}\right)f\left(\pi\frac{\sqrt{n^2+u}}{zk_m}\right)du\right.\\ &\left.-\int_0^{\infty}\int_0^{\infty}\left(\sqrt{n^2+u}\right)f\left(\pi\frac{\sqrt{n^2+u}}{zk_m}\right)dudn\right\}\end{aligned} \tag{3.30}$$

This form was amenable to exploiting the Euler–Maclaurin series,

$$\sum_{(0)1}^{\infty} F(n) - \int_0^{\infty} F(n)dn = -\frac{1}{12}F'(0) + \frac{1}{24 \times 30}F'''(0) + \cdots \quad (3.31)$$

He then turned Equation (3.30) into Equation (3.31), by making the substitution $w = u + n^2$, which allowed him to define $F(n)$ as,

$$F(n) = \int_{n^2}^{\infty} w^{1/2} f\left(\frac{w\pi}{ak_m}\right) dw \quad (3.32)$$

From this equation, the derivatives of $F(n)$ in the Euler–Maclaurin series were evaluated, to give: $F'(n) = -2n^2 f(n^2\pi/ak_m)$, so $F'(0) = 0$, and $F'''(0) = -4$. From this, the energy per plate area was found to be,

$$U_{Casimir}(z) = -\frac{\pi^2 \hbar c}{720}\frac{1}{z^3} \quad (3.33)$$

Finally, the Casimir force is evaluated as the gradient of the energy in the inter-plate direction, to give Casimir's formula,

$$\frac{F^0_{Cas}}{A} = -\frac{\pi^2 \hbar c}{240}\frac{1}{z^4} \quad (3.34)$$

Casimir gave this result the following interpretation: "There exists a force of attraction between two metal plates which is independent of the material of the plates as long as the distance is so large that for wavelengths comparable with that distance the penetration depth is small comparable with that distance. The force may be interpreted as the zero-point pressure of electromagnetic waves."

For planar parallel metallic plates with an area $A = 1\ \text{cm}^2$ separated by a distance $z = 0.5\ \mu\text{m}$, the Casimir force is $2 \times 10^{-6} N$.

3.1.3.2 Lifshitz' calculation of the casimir force

Lifshitz' approach was to calculate the molecular forces of attraction between two parallel solids separated by a medium of length l [50]. In this context, the problem could be approached in a purely macroscopic fashion, since the distance between bodies could be assumed to be large compared to the inter-atomic distance. The interaction of the bodies is then regarded as occurring by way of a fluctuating electromagnetic field. This field, in addition to being always present in the interior of any absorbing medium, also extends beyond

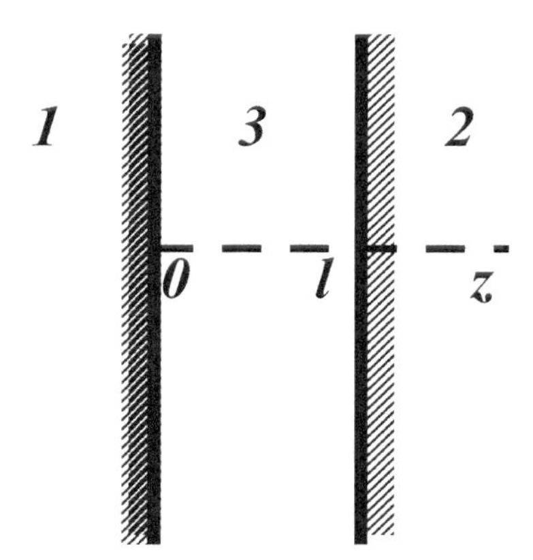

Figure 3.8 Lifshitz geometry. *1* and *2* represent solid bodies, *3* represents the medium between them, and *l* represents the distance between the bodies.

its boundaries in the form of traveling waves, radiated by the boundaries, and partially in the form of standing waves which are damped exponentially as one moves away from the surface of the body (Figure 3.8).

Lifshitz emphasized that this field does not vanish even at absolute zero, at which point it consists of the zero-point fluctuations of the radiation field. These facts endowed the analysis with a number of features. Firstly, it lends validity to the results at all temperatures; secondly, it takes into account retardation effects, which become prominent for large separation between the bodies; and thirdly, in the limiting case of rarefied media (bodies), it leads to the same results obtained by considering the interaction of individual atoms.

The first step in Lifshitz' calculation was to determine the fluctuating field [50]. This made use of Rytov's theory [51], which was based on introducing into Maxwell's equations a *random* field, as "forcing" function. In particular, according to this theory, in a dielectric, non-magnetic medium, these equations for a monochromatic field $\sim e^{-i\omega t}$ take the form,

$$\nabla \times \vec{E} = i\frac{\omega}{c}\vec{H} \tag{3.35}$$

$$\nabla \times \vec{H} = -i\frac{\omega}{c}\varepsilon\vec{E} - i\frac{\omega}{c}\vec{K} \tag{3.36}$$

where $\varepsilon(\omega)$ is the complex dielectric constant and K is the random field. The fundamental characteristic of K is the correlation function between components of K at two points in space, which is given by Rytov as,

$$K_i(x, y, z)K_k(x', y', z') = A\varepsilon''(\omega)\delta(x - x')\delta(y - y')\delta(z - z') \tag{3.37}$$

and

$$A = 4\hbar\left(\frac{1}{2} + \frac{1}{e^{\hbar\omega/T} - 1}\right) = 2\hbar\coth\frac{\hbar\omega}{2T} \tag{3.38}$$

where T is temperature and ε'' is the imaginary part of ε. The link to the problem geometry is captured by representing the function $K(x, y, z)$ by a Fourier integral of the form,

$$K(x, y, z) = \int_{-\infty}^{\infty} g(\vec{k})e^{i\vec{q}\cdot\vec{r}}\cos k_z z d\vec{k} \tag{3.39}$$

where q is taken as a two-dimensional vector with components k_x and k_y so that $k^2 = k_z^2 + q^2$, and r is the radius vector in the xy plane. In Equation (3.39), the Fourier coefficient $g(k)$ is given by the correlation function corresponding to the spatial correlation,

$$\overline{g_i(\vec{k})g_k(\vec{k}')} = \frac{A\varepsilon''}{4\pi^3}\delta_{ik}\delta(\vec{k} - \vec{k}') \tag{3.40}$$

Having found an explicit form for the random function K, Lifshitz solved Maxwell's equations subject to the boundary conditions imposed by the geometry (Figure 3.8) and then matched the solutions at the interfaces. In medium 1, the solution was found as,

$$\begin{aligned}\vec{E}_1 &= \int_{-\infty}^{\infty}\left\{\vec{a}_1(\vec{k})\cos k_z z + i\vec{b}_1(\vec{k})\sin k_z z\right\} e^{i\vec{q}\cdot r}d\vec{k} \\ &\quad + \int_{-\infty}^{\infty} \vec{u}_1(\vec{q})e^{i\vec{q}\cdot\vec{r} - is_1\cdot z}d\vec{q}\end{aligned} \tag{3.41}$$

and

$$\begin{aligned}\vec{H}_1 &= \frac{c}{\omega}\int_{-\infty}^{\infty}\{(\vec{q}\times\vec{a}_1 + k_z(\vec{n}\times\vec{b}_1)\cos k_z z) + i(\vec{q}\times\vec{b}_1) \\ &\quad + k_z(\vec{n}\times\vec{a}_1)\sin k_z z\}e^{i\vec{q}\cdot\vec{r}}d\vec{k} \\ &\quad + \frac{c}{\omega}\int_{-\infty}^{\infty}\{\vec{q}\times\vec{u}_1 - \vec{s_1}(\vec{n}\times\vec{u}_1)\}e^{i\vec{q}\cdot\vec{r} - is_1 z}d\vec{q}\end{aligned} \tag{3.42}$$

where **n** is a unit vector in the direction z, and

$$s_1 = \sqrt{\frac{\omega^2}{c^2}\varepsilon_1 - q^2} \tag{3.43}$$

the positive root is taken. The first terms of Equations (3.41) and (3.42) represent the inhomogeneous solution, that is, due to the forcing function K, with coefficients given by,

$$\vec{a}_1 = \frac{1}{\varepsilon_1\left(k^2 - \frac{\omega^2\varepsilon_1}{c^2}\right)}\left(\frac{\omega^2}{c^2}\varepsilon_1\vec{g}_1 - \vec{q}(\vec{q}\cdot\vec{g}_{1r}) - k_z^2 g_{1z}\vec{n}\right) \quad (3.44)$$

$$\vec{b}_1 = -\frac{k_z}{\varepsilon_1\left(k^2 - \frac{\omega^2\varepsilon_1}{c^2}\right)}\left[\vec{n}(\vec{q}\cdot\vec{g}_{1r}) + q g_{1r}\right] \quad (3.45)$$

where two-dimensional vectors in the xy plane have been given the subscript r. The second integrals in Equations (3.41) and (3.42) represent the homogeneous solution to Maxwell's equations (i.e., $K = 0$) and are interpreted as describing the plane wave reflected from the boundary of the medium, with the condition for transversality of these waves given by,

$$\vec{u}_{1r}\cdot\vec{q} - s_1 u_{1z} = 0 \quad (3.46)$$

The solution in medium 2 are identical to those in medium 1, except that they are "displaced," by replacing $cos\, k_z z, sin\, k_z z$ by $cos\, k_z(z-l), sin\, k_z(z-l)$ and changing the sign of s, the reflected waves, which now propagate in the positive z direction.

The solutions in the space between the bodies, 3, (Figure 3.8) are found by assuming $\varepsilon = 1$ and $K = 0$ and have the form,

$$E_3 = \int_{-\infty}^{\infty}\left\{\vec{v}(\vec{q})e^{ipz} + \vec{w}(\vec{q})e^{-ipz}\right\}e^{i\vec{q}\cdot\vec{r}}d\vec{q} \quad (3.47)$$

$$\begin{aligned}\vec{H}_3 = {} & \frac{c}{\omega}\int_{-\infty}^{\infty}\left\{\vec{q}\times\vec{v} + \vec{p}(\vec{n}\times\vec{v})e^{ipz} + (\vec{q}\times\vec{w})\right. \\ & \left. -p(\vec{n}\times\vec{w})e^{-ipz}\right\}e^{i\vec{q}\cdot\vec{r}r}d\vec{q}\end{aligned} \quad (3.48)$$

here,

$$p = \sqrt{\frac{\omega^2}{c^2} - q^2} \quad (3.49)$$

and v and w satisfy the transversality conditions,

$$\vec{v}_r\cdot\vec{q} + p v_z = 0, \quad \vec{w}_r\cdot\vec{q} - p w_z \quad (3.50)$$

The formulation of the problem is completed by specifying the boundary conditions as continuity of **E** and **H** at the interfaces $z = 0$ and $z = l$. At $z = 0$, they take the form,

$$\int_{-\infty}^{\infty} a_{1r} dk_z + \vec{u}_{1r} = \vec{v}_r + \vec{w}_r \tag{3.51}$$

$$\int_{-\infty}^{\infty} (\vec{q}a_{1z} - k_z b_{1r}) dk_z + \vec{q}u_{1z} + s_1 u_{1z} + s_1 u_{1r} = \vec{q}(v_z + w_x) - \vec{p}(v_r - w_r) \tag{3.52}$$

At $z = l$, they take the same form, except that $\mathrm{s}_1, \mathrm{a}_1, \mathrm{b}_1$, **v**, and **w** are replaced by $\mathrm{s}_2, \mathrm{a}_2, \mathrm{b}_2, ve^{ipl}$, and $\mathrm{w}e^{-ipt}$, respectively.

Solution of the Maxwell's equations subject to the boundary and continuity conditions yields all the field amplitudes, where the quantity q runs through values from zero to infinity, and is interpreted as corresponding to undamped plane waves in the region 3, between the two bodies, while p runs through real values from w/c to zero, and purely imaginary values from zero to $i\infty$, and is interpreted as referring to exponentially damped plane waves.

The force of attraction F between the two bodies is then calculated as the zz component of the Maxwell stress tensor. This component is expressed as,

$$F = \int_0^{\infty} F_\omega d\omega = \frac{1}{4\pi} \int_0^{\infty} \left\{ \overline{E}_{3r}^2 + \overline{H}_{3r}^2 - \overline{E}_{3z}^2 - \overline{H}_{3z}^2 \right\}_{z=0} d\omega \tag{3.53}$$

where the bar over a symbol signifies a statistical averaging to which the Fourier component g of the random field must be subjected. What follows in Lifshitz' derivation is a *tour de force* in the analytical evaluation of the integral. After a number of transformations, he manages to express F_w as,

$$F_\omega = \frac{\hbar}{4\pi^2} \coth \frac{\hbar\omega}{2T} \times \int p^2 dp \left\{ \left[\frac{(s_1 + p)(s_2 + p)}{(s_1 - p)(s_2 - p)} e^{-2ipl} - 1 \right]^{-1} + \left[\frac{(s_1 + \varepsilon_1 p)(s_2 + \varepsilon_2 p)}{(s_1 - \varepsilon_1 p)(s_2 - \varepsilon_2 p)} e^{-2ipl} - 1 \right]^{-1} + \frac{1}{2} \right\} + c.c. \tag{3.54}$$

where c.c. means the complex conjugate of the first term, and the integration over p is to be carried out in the plane of the complex variable p, over the segment $(w/c, 0)$ of the real axis and over the upper half of the imaginary axis. While the above formula captures the monochromatic Maxwell stress tensor F_w, at any temperature, it is the following $T = 0$ formula that has become the hallmark of Lifshitz' calculation [50, 52],

$$F = \frac{\hbar}{2\pi^2 c^3} \times \int_0^\infty \int_1^\infty p^2 \xi^3 \left\{ \left[\frac{(s_1+p)(s_2+p)}{(s_1-p)(s_2-p)} e^{2p\xi l/c} - 1 \right]^{-1} \right.$$
$$\left. + \left[\frac{(s_1+p\varepsilon_1)(s_2+p\varepsilon_2)}{(s_1-p\varepsilon_1)(s_2-p\varepsilon_2)} e^{2p\xi l/c} - 1 \right]^{-1} \right\} dp d\xi \tag{3.55}$$

where $\omega = i\xi$ for imaginary values of ω, and ε_1 and ε_2 are to be taken as real functions of $\varepsilon_1(i\xi)$ and $\varepsilon_2(i\xi)$. This formula makes it possible to compute the force F for any separation l and materials, including dielectrics, if the functions $\varepsilon(i\xi)$ are known for both bodies. This latter function can be expressed as,

$$\varepsilon(i\xi) - 1 = \frac{2}{\pi} \int_0^\infty \frac{\omega \varepsilon''(\omega)}{\omega^2 + \xi^2} d\omega \tag{3.56}$$

The crux of Lifshitz' approach to the above calculations is the determination of electromagnetic field between the bodies and the computation of the corresponding Maxwell stress tensor [50, 52].

3.1.3.3 Casimir force calculation of brown and maclay

Brown and Maclay [53] calculated the Casimir force between two bodies (conducting parallel plates) by obtaining the stress–energy tensor $T^{\mu v}(x)$ from the definition,

$$T^{\mu v}(x) = \lim_{\varepsilon \to 0} \left(1 + \frac{1}{4} \varepsilon^\lambda \frac{\partial}{\partial \varepsilon^\lambda} \right) T^{\mu v}(x, \varepsilon) \tag{3.57}$$

where

$$T^{\mu v}(x, \varepsilon) = F^{\mu\lambda} \left(x + \frac{1}{2}\varepsilon \right) F^v \lambda \left(x - \frac{1}{2} \right)$$
$$- \frac{1}{4} g^{\mu v} F^{\lambda k} \left(x + \frac{1}{2} \right) F_{\lambda k} \left(x - \frac{1}{2} \right) \tag{3.58}$$

is the quantum electrodynamic stress tensor. In particular, the stress–energy tensor was defined so that infinite quantities never appeared, and it was explicitly computed with the aid of an image-source construction of the Green's function.

Considering first the situation at zero temperature, the Green's function was constructed with an infinite sequence of current-pulse image sources displaced along the z-axis, but which exist at a common infinitesimal time duration. Due to the special symmetry of the parallel-plate geometry, which has pairs of sources at equal distances from a given plate, no retardation was required for their radiation pulses to reach the plates simultaneously. The constructed Green's function is,

$$\begin{aligned} G_{+}^{\mu v;\lambda\kappa}(x,x') &= \langle iT^{*}(F^{\mu v}(x))F^{\lambda\kappa}(x')\rangle_{\langle 0\rangle} \\ &= d^{\mu v;\lambda\kappa}\sum_{l=-\infty}^{\infty} D_{+}(x-x'-2al\hat{z}) - \overline{d}^{\mu v;\lambda\kappa} \\ &\qquad \sum_{l=-\infty}^{\infty} D_{+}(x-\overline{x}'-2al\hat{z}) \end{aligned} \tag{3.59}$$

where $\hat{z}^{\mu} = (0,0,0,1)$ is a four vector and $g^{\mu v}$ is the metric tensor with signature, $(-1,1,1,1)$,

$$D_{+}^{\mu v;\lambda\kappa}(x-x') = d^{\mu v;\lambda\kappa}D_{+}(x-x') \tag{3.60}$$

$$d^{\mu v;\lambda\kappa} = \partial^{\mu}\partial'^{\lambda}g^{v\kappa} - \partial^{v}\partial'^{\lambda}g^{\mu\kappa} + \partial^{v}\partial'^{\kappa}g^{\mu\lambda} - \partial^{\mu}\partial'^{\kappa}g^{v\lambda} \tag{3.61}$$

and the zero-mass propagator is given by,

$$D_{+}(x) = i\int \frac{d\vec{k}}{(2\pi)^{3}}\frac{1}{2|k|}e^{i\vec{k}.\vec{r}-i|k||t|} = \frac{i}{4\pi^{2}}\frac{1}{x^{2}+i\varepsilon} \tag{3.62}$$

Each term in the sum corresponds to a particular reflection of the original source pulse by one of the plates, and since an infinite number of such reflections is possible, the sums contain an infinite number of image terms. Then, the stress tensor is given by,

$$\begin{aligned} \langle T^{\mu v}\rangle &= (-i)G^{\mu\lambda;v}\lambda(x,x) - \frac{1}{4}g^{\mu v}(-i)G^{\lambda\kappa};\lambda\kappa(x,x) \\ &= -\partial^{\mu}\partial^{v}\sum_{l=-\infty}^{\infty}(-i)D_{+}(x-x'-2al\hat{z})|_{z=z'} \end{aligned} \tag{3.63}$$

where according to Equation (3.63), it is implicit that the vacuum contribution to the Green's function is omitted, so that the value $l = 0$ is excluded from the sum. It turns out that due to the massless nature of the photon, the theory contains no intrinsic unit of length and is invariant under scale transformation of the electromagnetic field strength. Consequently, this invariance is manifested in the vanishing of the trace of the stress–energy tensor, that is,

$$T^{\mu}_{\mu}(x) = 0 \tag{3.64}$$

Under these circumstances, Brown and Maclay [53] find that,

$$\langle T^{\mu v} \rangle_{\langle 0 \rangle} = \left(\frac{1}{4} g^{\mu v} - \hat{z}^{\mu} \hat{z}^{v} \right) (\hbar c / a^4) \gamma \tag{3.65}$$

in which γ is a pure number, with numerical value,

$$\gamma = \frac{1}{2\pi^2} \sum_{l=1}^{\infty} l^{-4} = \frac{1}{2\pi^2} \zeta(4) = \frac{\pi^2}{180} \tag{3.66}$$

This results in the energy density between the plates,

$$\langle T^{00} \rangle_{\langle 0 \rangle} = -\frac{1}{4} (\hbar c / a^4) \gamma = -\left(\frac{\pi^2}{720} \right) \cdot \left(\frac{\hbar c}{a^4} \right) \tag{3.67}$$

and the pressure between the plates as,

$$\langle T^{33} \rangle_{\langle 0 \rangle} = -\frac{3}{4} (\hbar c / a^4) \gamma = -\left(\frac{\pi^2}{240} \right) \cdot \left(\frac{\hbar c}{a^4} \right) \tag{3.68}$$

which is the Casimir force result.

3.1.3.4 Casimir force calculations for arbitrary geometries

Due to its impact in a wide variety of devices, there is a strong interest in schemes for computing Casimir forces in systems with arbitrary geometries. This was the aim of the works advanced by Emig et al. [54], Milton and Wagner [55], Reid et al. [56], Rodriguez et al. [57], McCauley et al. [58], and Babington and Scheel [59].

3.1.3.4.1 *Computing the casimir energy based on multipole interactions*

The multipole expansion approach has been recently applied by Emig et al. [54], as an exact way for computing the Casimir energy between arbitrary compact objects, either dielectrics or perfect conductors. The development, which is not new, as indicated by Milton and Wagner [55], entailed computing the energy as an interaction between multipoles generated by quantum current fluctuations. In particular, the objects' shape and composition was captured only through their scattering matrices. In this context, the Casimir energy was given by [54],

$$E = \frac{\hbar c}{2\pi} \int_0^{\infty} d\kappa \log \det(I - U^{-}T^{2}U^{+}T^{1}) \tag{3.69}$$

in terms of universal matrices U^{-} and U^{+}, representing the interaction between multipoles, and T-matrices, related to the scattering matrices of the objects 1 and 2 and capturing their shape and material properties. Examples of the Casimir energy between two dielectric spheres and the full interaction at all separations for perfectly conducting spheres were given. In principle, the method was claimed to be applicable to dielectric objects of any shape, whose T matrix can be obtained by integrating the standard vector solutions of the Helmholtz equation in dielectric media over the object's surface. In addition, both the analytical and numerical results that are available for many shapes can be exploited.

An efficient algorithm, following the approach being discussed, was recently applied to perfectly conducting non-spheroidal, non-axisymmetric objects and objects with sharp corners [56]. The algorithm departs from,

$$E = -\frac{\hbar c}{2\pi} \int_0^{\infty} d\kappa \log \frac{Z(\kappa)}{Z_{\infty}(\kappa)} \tag{3.70}$$

with,

$$Z(\kappa) = \int D\vec{J}(\vec{x}) e^{-(1/2)\int dx \int d\vec{x}' \vec{J}(\vec{x}) \cdot G_x(\vec{x}, \vec{x}') \cdot \vec{J}(\vec{x}')} \tag{3.71}$$

where the functional integration extends over all possible current distributions $J(x)$ on the surfaces of the objects and where $G_x = \left[1 + \frac{1}{\kappa^2} \nabla \otimes \nabla'\right] \frac{e^{-\kappa|x-x'|}}{4\pi|x-x'|}$ is the dyadic Green's function at frequency $i\omega = c\kappa . Z_{\infty}$ is Z computed with all objects removed to infinite separation [55]. A computationally tractable version of Equation (3.71) was developed by expressing the current distribution in a discrete basis, $\vec{J}(\vec{x}) = \sum J_{im} \vec{f}_{im}(\vec{x})$, where $i = 1, \ldots, N_0$ ranges over the objects in the geometry and $m = 1, \ldots, N_i$ ranges over a set of N_i expansion functions defined for the ith object. This resulted in the below equation,

$$E = +\frac{\hbar c}{2\pi} \int_0^{\infty} d\kappa \log \frac{\det M(\kappa)}{\det M_{\infty}(\kappa)} \tag{3.72}$$

where the matrix elements of $\mathbf{M}(\kappa)$ are the interactions between the basic functions,

$$M_{\alpha\beta}(\kappa) = \int\int \vec{f}_\alpha \cdot G_\kappa \cdot \vec{f}_\beta d\vec{x} d\vec{x}' \tag{3.73}$$

The z-directed Casimir force on the ith object is given by [56],

$$F_{z,i} = -\frac{\hbar c}{2\pi}\int_0^\infty d\kappa \frac{\partial}{\partial z_i} \ln \frac{\det M(\kappa)}{\det M_\infty(\kappa)} \tag{3.74}$$

3.1.3.4.2 *Computing the casimir force using finite-difference time-domain techniques*

For engineering applications, techniques that are familiar to the engineering community are essential. This is the case of a method recently introduced by Rodriguez et al. [57, 58] to compute Casimir forces in arbitrary geometries and for arbitrary materials based on the FDTD approach. This method captures the time evolution of electric and magnetic fields in response to a set of current sources in a modified medium with frequency-independent conductivity. In contrast to the above method, this approach has the advantage that it allows the exploitation of existing FDTD software, without modification, to compute Casimir forces.

The method exploits the fact that the Casimir force on a body enclosed by any closed surface S can be expressed [60] as an integral over S of the mean electromagnetic stress tensor $\langle T_{ij}(\vec{r},\omega)\rangle$, where r denotes spatial position and ω frequency. The electromagnetic stress tensor, in turn, is related to the Casimir force by Ref. [57]. The details of the numerical algorithm for computing the Casimir force, rather convoluted, are given by,

$$F_i = \int_0^\infty d\omega \oint_S \sum_j \langle T_{ij}(\vec{r},\omega)\rangle dS_j \tag{3.75}$$

where i is the direction of the force. To compute the stress tensor, it is expressed in terms of the correlation functions $\langle E_i(\vec{r},\omega)E_j(\vec{r}',\omega)\rangle$ and $\langle H_i(\vec{r},\omega)H_j(\vec{r}',\omega)\rangle$ by,

$$\begin{aligned}\langle T_{ij}(\vec{r},\omega)\rangle = {} & \mu(\vec{r},\omega)\left[\langle H_i(\vec{r})H_j(\vec{r})\rangle_\omega - \frac{1}{2}\delta_{ij}\sum_k \langle H_k(\vec{r})H_k(\vec{r})\rangle_\omega\right] \\ & + \varepsilon(\vec{r},\omega)\left[\langle E_i(\vec{r})E_j(\vec{r})\rangle_\omega - \frac{1}{2}\delta_{ij}\sum_k \langle E_k(\vec{r})E_k(\vec{r})\rangle_\omega\right]\end{aligned} \tag{3.76}$$

Expressing both the electric and magnetic field correlation functions as derivatives of a vector potential operator $\vec{A}^E(\vec{r},\omega)$, defined as,

$$E_i(\vec{r},\omega) = -i\omega A_i^E(\vec{r},\omega) \tag{3.77}$$

$$\mu H_i(\vec{r},\omega) = (\nabla\times)_{ij} A_j^E(\vec{r},\omega) \tag{3.78}$$

in which the superscript indicates that **E** is obtained as a time-derivative of **A**, it is possible to exploit the fluctuation–dissipation theorem to relate the correlation function of $\mathbf{A}^\mathbf{E}$ to the photon Green's function $G_{ij}^E(\omega;\vec{r},\vec{r}\,')$, as,

$$\langle A_i^E(\vec{r},\omega)A_j^E(\vec{r}\,',\omega)\rangle = -\frac{\hbar}{\pi}\text{Im } G_{ij}^E(\omega,\vec{r},\vec{r}\,') \tag{3.79}$$

In Equation (3.79), $G_{ij}^E(\omega,\vec{r},\vec{r}\,')$ represents the vector potential A_i^E in response to an electric dipole current J along the $\widehat{e}_j$ direction, which is obtained from the solution to the equation,

$$\left[\nabla\times\frac{1}{\mu(\vec{r},\omega)}\nabla\times -\omega^2\varepsilon(\vec{r},\omega)\right] G_j^E(\omega;\vec{r},\vec{r}\,') = \delta(\vec{r}-\vec{r}\,')\widehat{e}_j \tag{3.80}$$

Once G_{ij}^E is obtained, Equations (3.77) and (3.78) are combined with Equation (3.79) to express the field correlation functions at points **r** and **r**$'$ in terms of the photo Green's function, to obtain,

$$\langle E_i(\vec{r},\omega)E_j(\vec{r}\,',\omega)\rangle = \frac{\hbar}{\pi}\omega^2 \text{ Im } G_{ij}^E(\omega,\vec{r},\vec{r}\,') \tag{3.81}$$

as expressed in Ref. [57]. The approach, once implemented, has been applied to a number of unusual geometries (Figure 3.9).

3.1.3.4.3 *Computing the casimir force using the framework of macroscopic quantum electrodynamics*

Since it is expected that future nanoelectromechanical quantum circuits and systems will exploit economies of scale, that is, to be dense systems, it is essential to develop approaches to computing the Casimir force in the context of configurations of multiple objects. This, in fact, was the aim of a recently presented approach by Babington and Scheel [59], where they developed an expression for the general Casimir force in an N-sphere system (Figure 3.10).

The approach employed the canonical stress tensor to calculate the resultant force on one of the spheres in the configuration. The stress–energy tensor

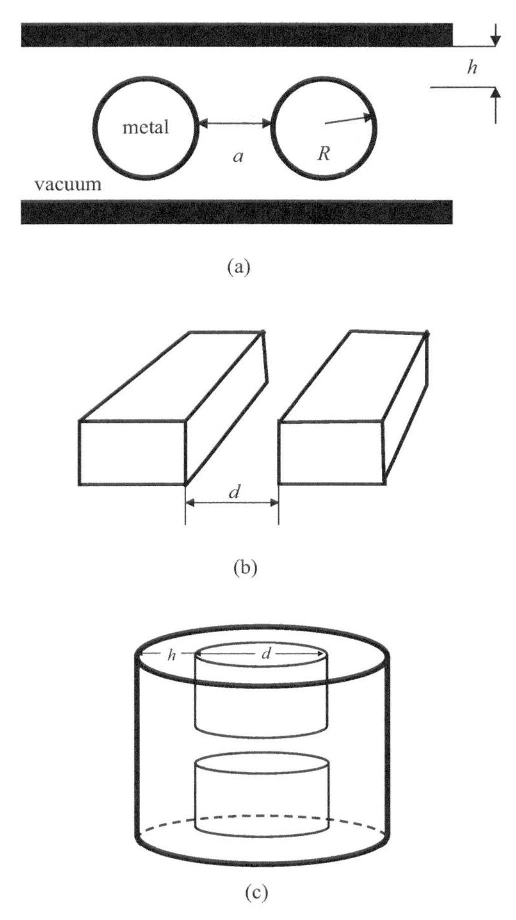

Figure 3.9 Nontrivial geometries for which Casimir forces were calculated using the FDTD method. (a) Parallel cylinders, (b) long silicon waveguides suspended in air, and (c) cylindrically symmetric piston.

utilized was the standard vacuum expression (which is consistent with the Lorentz force law), given by [59],

$$T_{ij}(x) = E_i(x)E_j(x) + B_i(x)B_j(x) - \frac{1}{2}\delta_{ij}\left(|E(x)|^2 + |B(x)|^2\right) \quad (3.82)$$

where it is assumed that the following limits for the initial and final points are being taken,

$$\lim_{x_1 \to x_2} E(x_1)E(x_2) = E(x_2)E(x_2) \quad (3.83)$$

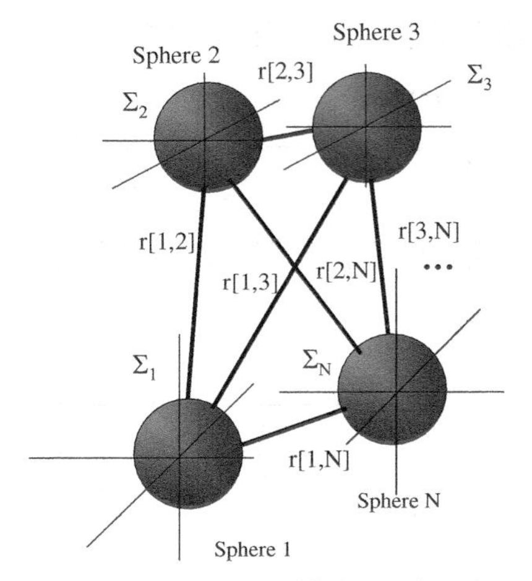

Figure 3.10 The N-Sphere system consists of N dielectric spheres of radii R[1], ..., R[N] each centered on N separate coordinate systems $\sum 1, \ldots, \sum$ N, all contained in a background dielectric.

$$\lim_{x_1 \to x_2} B(x_1)B(x_2) = B(x_2)B(x_2) \tag{3.84}$$

from which the scattering correlation functions for the electric field are evaluated as,

$$\lim_{y \to x} E_i(x)E_j(y) = \int_0^\infty d\omega d\omega' \langle E_i^{out}(x;\omega)^+ E_j^{in}(y;\omega')\rangle \tag{3.85}$$

and similarly for magnetic fields. The force on sphere 1, exerted by the other $N-1$ spheres, is then given by,

$$F_j(1|N-1) = \int_{B^2} d^3x \nabla_i T_{ij}(x) \tag{3.86}$$

where the volume B^2 is the ball that has the two-dimensional sphere as its boundary.

In calculating Equation (3.86), a multiple scattering approach was used to determine the fields on the sphere where the force was being determined, as a function of the scattered fields in all the other spheres. This entailed constructing the scattering two-point function by writing the fields in a mode decomposition of spherical vector wave functions, in which the "in" and "internal" states are regular at the ith-sphere origin, while the "out" states are outgoing modes falling off at infinity. In particular, they are eigenfunction modes with respect to the ith-sphere of the radial eigenfunctions, represented by spherical Bessel and Hankel functions. The approach was demonstrated by evaluating the Casimir force between two and three spheres at zero and finite temperature.

3.1.3.5 Corrections to ideal casimir force derivation

In practice, of course, the shape of material bodies may neither be cubical nor be perfectly smooth, and their conductivity is not infinite. Therefore, many corrections for the Casimir force, derived from experiments that measure it under various conditions, such as effecting normal displacement between a sphere and a smooth planar metal and between parallel metallic surfaces, as well as, effecting lateral displacement between a sphere and a sinusoidally corrugated surface, have been performed [39–43].

Corrections to the ideal expression (Equation 3.28) have been introduced to account for certain deviations. For example, for the sphere-plate geometry, the zero-temperature Casimir force is given by,

$$F^0_{Cas\_Sphere-Plate}(z) = -\frac{\pi^3}{360} R \frac{\hbar c}{z^3} \tag{3.87}$$

where R is the radius of curvature of the spherical surface.

Similarly, to include the finite conductivity of the metallic boundaries, two approaches have been advanced. In the first one, the force is modified as [44, 45],

$$F^{0,\sigma}_{Cas}(z) = F^0_{Cas\_Sphere-Plate}(z)\left[1 - 4\frac{c}{z\omega_p} + \frac{72}{5}\left(\frac{c}{z\omega_p}\right)^2\right] \tag{3.88}$$

where ω_p is the metal plasma frequency [46]. In the second one, obtained by Lifshitz [47], the correction is ingrained in the derivation of the Casimir force and is given by,

$$\begin{aligned} F^{0,\sigma}_{Cas}(z) = -\frac{R\hbar}{\pi c^3}\int_0^z dz' \int_0^\infty \int_1^\infty p^2\xi^3 dpd\,\xi \\ \times \left\{ \begin{array}{l} \left[\frac{(s+p)^2}{(s-p)^2} e^{\frac{2p\xi z}{c}} - 1\right]^{-1} \\ + \left[\frac{(s+p\varepsilon)^2}{(s-p\varepsilon)^2} e^{\frac{2p\xi z}{c}} - 1\right]^{-1} \end{array} \right\} \end{aligned} \tag{3.89}$$

where $s = \sqrt{\varepsilon - 1 + p^2}, \varepsilon(i\xi) = 1 + \frac{2}{\pi}\int_0^\infty \frac{\omega\varepsilon''(i\xi)}{\omega^2+\xi^2} d\omega$ is the dielectric constant of the metal, ε'' is the imaginary component of ε, and ξ is the imaginary frequency given by $\omega = i\xi$.

Corrections due to nonzero temperature yield [28],

$$F_{Cas}^{T}(z) = F_{Cas}^{0}(z)\left[1 + \frac{720}{\pi^2} f(\zeta)\right] \tag{3.90}$$

where $\zeta = k_B T z/\hbar c, k_L$ is the Boltzmann constant, T is the absolute temperature, and

$$f(\zeta) \approx \begin{cases} (\zeta^3/2\pi)\vartheta(3) - (\zeta^4\pi^2/45), & \text{for} \quad \zeta \leq 1/2 \\ (\zeta/8\pi)\vartheta(3) - (\pi^2/720), & \text{for} \quad \zeta > 1/2 \end{cases} \tag{3.91}$$

with $\vartheta(3) = 1.202\ldots$.

Roy and Mohideen [48] included the effects of surface roughness, which changes the surface separation, by replacing the flat plate with a spatial sinusoidal modulation of period λ, and the energy averaged over the size of the plates, L, to obtain,

$$\left\langle U_{Casimir}\left(z + A\sin\frac{2\pi x}{\lambda}\right)\right\rangle = -\frac{\pi^2\hbar c}{720}\frac{1}{z^3}\sum_m C_m\left(\frac{A}{z}\right)^m \tag{3.92}$$

where A is the corrugation amplitude. The corresponding Casimir force is given by the so-called *force proximity theorem* [49] relating the parallel-plate geometry and the sphere-plate geometry,

$$F_{Cas\_Roughness} = 2\pi R\langle U_{Cas\_Roughness}\rangle \tag{3.93}$$

For $\lambda \ll L$ and $z + z_0 > A$, where z_0 is the average surface separation after contact due to stochastic roughness of the metal coating, they suggest the following coefficients in Equation (3.92): $C_0 = 1, C_2 = 3, C_4 = 45/8, C_6 = 35/4$. A more accurate and general model for stochastic surface roughness, advanced by Harris et al. [36], includes the effects of surface roughness, by replacing the flat plate with the mean stochastic roughness amplitude A, to obtain,

$$F_{Cas}^{r}(z) = F_{Cas}^{0}(z)\left[1 + 6\left(\frac{A}{z}\right)^2\right] \tag{3.94}$$

where A is derived from direct measurements via an atomic force microscope (AFM).

3.1.4 Radiation Pressure Actuation

As is well known, electromagnetic (EM) radiation carries energy as it propagates [17]. In particular, the energy flux Watts/m^2 transported by an EM wave propagating in free space is given by the Poynting vector (Equation 3.95),

$$\vec{S} = \frac{1}{\mu_0} \vec{E} \times \vec{B} \tag{3.95}$$

where $\bar{E}$ and $\bar{B}$ are the instantaneous electric and magnetic field vectors, respectively, and μ_0 is the permeability of free space. The portion of this energy, U, that is totally absorbed by an object on which the wave impinges, exerts a pressure on the object, called *radiation pressure*, manifested as the transfer of a momentum of magnitude, p, given by [17].

$$p = \frac{U}{c} \tag{3.96}$$

where c is the speed of light. This momentum is experienced by the object in the direction of the incident light beam. On the other extreme, when the light is totally reflected by the object, the amount of momentum exerted on the object is *twice* that given in Equation (3.96) [17].

A planar mirror, with 100% reflection, having an area $A = 1$ cm^2, and on which a parallel light beam with an energy flux $S = 5$ Watts/cm^2 impinges for a period t of 1 hour, will reflect an energy given by,

$$U = (5 \text{ Watts/cm}^2) \times (1 \text{ cm}^2) \times (3600 \text{ sec}) = 1.8 \times 10^4 \text{ Joules} \tag{3.97}$$

and experience a momentum given by,

$$p = \frac{2U}{c} = \frac{2 \times 1.8 \times 10^4 \text{ Joules}}{3 \times 10^8 \text{ meters/sec}} = 1.2 \times 10^{-4}\text{kg-m/sec} \tag{3.98}$$

This is equal to experiencing an average force,

$$F = \frac{p}{t} = \frac{1.2 \times 10^{-4}\text{kg-m/sec}}{3600 \text{ sec}} = 3.33 \times 10^{-8}\text{N} \tag{3.99}$$

This force is, clearly, too small to be of noticeable or of consequence to macroscopic objects, like human beings.

When the light beam's power impinging upon an object is set to P, and the reflection from the object is neither zero nor one, say a quantity, $0 \leq q \leq 1$, then the radiation force on the object takes the form [62],

$$F = \frac{2qP}{c} \tag{3.100}$$

In the context of submicron and atomic scale objects, the radiation pressure-derived force was employed in 1970 to accelerate and trap atomic particles [62]. Ashkin, the scientist who performed such experiments for the first time, was honored with a Nobel Prize in Physics in 2018. Next, we review some aspects of Ashkin's original work.

3.1.4.1 Radiation pressure manipulation of particles

An experiment demonstrating the manipulation of spherical transparent latex particles of diameters $0.59\ \mu m$, $1.31 \mu m$, and $2.68\ \mu m$, freely suspended in water, was carried out by Ashkin [62] (Figure 3.11).

The experimental setup (Figure 3.11) utilized as the light source the TEM_{00}-mode beam of an argon laser of radius $W_0 = 6.2\ \mu m$ and wavelength $\lambda = 0.514\ \mu m$, which was focused horizontally through a glass cell with a thickness of $120\ \mu m$, and manipulated to focus on single particles. Ashkin observed, in particular, that a beam with milliwatts of power impinging off center upon a $2.68\ \mu m$ sphere caused it to be simultaneously drawn into the beam axis and accelerated in the direction of the light propagation. These two particle displacements, namely one drawing the particles simultaneously into the beam axis and along the beam direction of propagation are the result of forces that derive from the power distribution profile and the refraction coefficient difference outside and inside the spheres (Figure 3.12). With the beam maximum power being along the A-axis, the strength of the beam along a is greater than that along b. Therefore, as seen, the force components closer to a, namely the deflected light beam forces, F_D^i and F_D^0, are much larger than those closer to b, namely the refracted light beam forces, F_R^i and F_R^0. Thus, in terms of vectorial components, it is seen that the vector sum of F_D^i and F_D^0 in the $-\mathrm{r}$ direction adds, thereby pulling the sphere upward towards the A-axis where the larger part of the light intensity lies. Similarly,

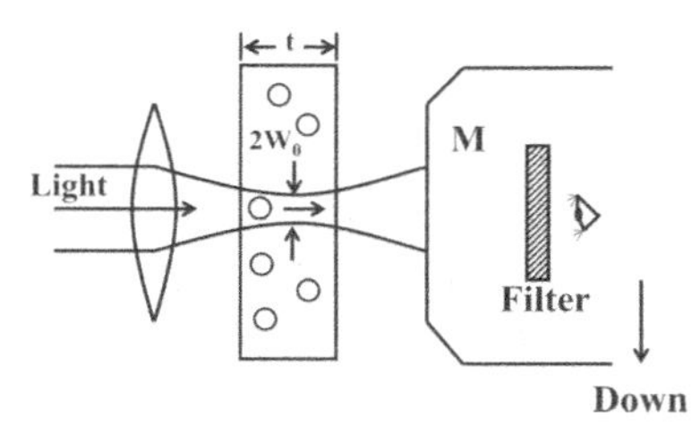

Figure 3.11 Geometry of glass cell, $t = 120\ \mu m$, for observing micron particle motions in a focused laser beam with a microscope M.

Source: Ref. [62].

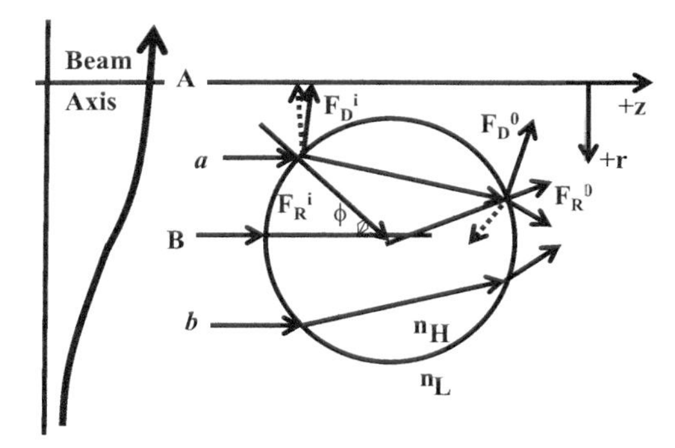

Figure 3.12 A dielectric sphere situated off the A-axis of a TEM_{00}-mode beam and a pair of symmetric rays a and b. The forces due to a are shown for $n_H = 1.58$ and $n_L = 1.33$. The sphere moves toward +z and −r.

Source: Ref. [62].

the +z components of F_D^i and F_D^0 add, producing an acceleration of the sphere in the $+z$ direction. On the other hand, the radial force components inside the sphere, F_R^i and F_R^0, accompanying the refracted light tend to cancel.

As a result of the acceleration experienced by the particles due to the radiation force, they acquire a velocity v in the water. This velocity is given by Stokes' law [51],

$$v = \frac{2q\,\mathrm{Pr}}{3c\pi W_0^2 \eta} \tag{3.101}$$

where η is the viscosity of the medium, $r \ll W_0$. An idea of the magnitude of v may be obtained by considering a light beam with power $P = 19\ \mathrm{mW}$ and beam radius of $6.2\ \mu\mathrm{m}$, impinging on a sphere of radius $1.34\ \mu\mathrm{m}$ and $q = 0.06$, submerged in water ($\eta = 10^{-2}P$). Then, following Equation (3.101), one obtains a velocity $v = 29\ \mu\mathrm{m/sec}$.

3.1.4.2 Radiation pressure trapping of particles

The physics described above pertaining to radiation pressure-induced forces may be exploited to "trap" or suspend particles is a state of zero velocity; this was the contribution that won Ashkin the Physics Nobel Prize in 2018. He accomplished this feat by way of the arrangement shown in Figure 3.13.

Ashkin's reasoning behind his successful experiment was as follows [62]: "If one has two opposing equal TEM_{00} Gaussian beams with beam waists located as shown in Figure 3.10, then a sphere of high index will be in stable

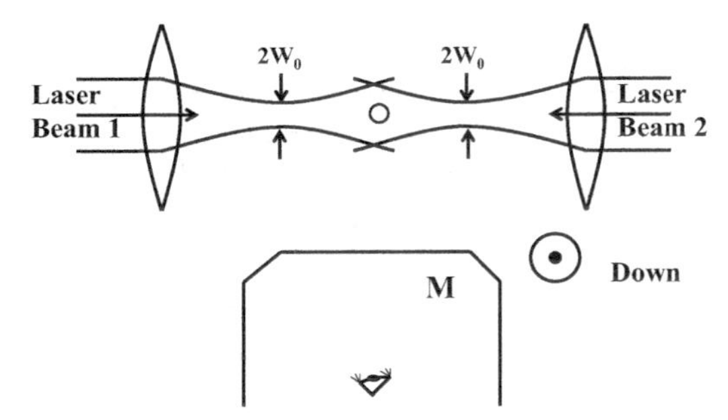

Figure 3.13 The trapping of a high index particle in a stable optical well produced by counter propagating light beams. Observation of the particle through the microscope M verifies its static nature.

Source: Ref. [62].

equilibrium at the symmetry point as shown (i.e., any displacement gives a restoring force)." It was observed that particles that drift in the neighborhood of the beams are drawn in toward the maximum power region and accelerated to a stable point at which they stop.

It is curious to note that if one of the opposing beams is interrupted, then the particle moves in the direction of the remaining beam, and that if the remaining beam is also interrupted or blocked, then the particle moves at random according to Brownian motion [62].

3.1.4.3 Radiation pressure effect on cantilever beams

The observation of radiation pressure forces on particles is normally hampered by the radiation (laser) power-induced heating of the particles as they absorb part of the radiation energy [62]. In fact, heating effects, manifested as a tendency of the particles to move in response to the establishment of a temperature gradient surrounding them, usually obscures radiation pressure forces. Ashkin avoided this scenario by utilizing highly transparent particles in a highly transparent medium, namely water [62].

On the other hand, the effects of radiation-induced heating are more difficult to counter in the case of a cantilever beam, where the absorbed optical radiation power may significantly heat up the cantilever beam causing it to deflect as a result of differential (non-uniform) expansion [18, 63]. This may be visualized when the radiation force is expressed as,

$$F = \frac{(2P_R + P_A)}{c} \tag{3.102}$$

where P_R and P_A are the reflected and absorbed radiation powers, respectively [63]. In this context, a technique to enhance the beam response to the optical radiation, which exploits that modulating the optical radiation at a frequency high enough that it minimizes the heating effects, has been advanced [63].

For a cantilever beam of length L and spring constant k, with a continuous laser beam spot impinging at a distance x from the anchor and causing a reflected power P_R and an absorbed power P_A, the resulting displacement at the beam tip is given by [63],

$$\delta^{rad} = \frac{(2P_R + P_A)(a^2 - 3a + 3)}{3cka(a-2)^2} \tag{3.103}$$

where $a = x/L$ and c is the speed of light. Under laser beam amplitude modulation conditions, the resulting root-mean-square (RMS) beam tip displacement is given by [63],

$$\delta^{rad}_{RMS} = \frac{1}{2\sqrt{2}} \frac{(2P_R + P_A)(a^2 - 3a + 3)}{3cka(a-2)^2} \tag{3.104}$$

The properties of radiation pressure-driven cantilever beams have been studied experimentally by Evans et al. [63] and Ma, Garrett, and Munday [64]. The conceptual setup is captured by the sketch in Figure 3.14. Here, a laser beam is applied to the lower beam surface (from underneath), and the deflection of another laser beam impinging on the top beam surface is measured and processed to extract the oscillation amplitude of the beam. It is found that the beam vibration spectral density versus frequency depends on the relationship between the rates at which the beam heats up and cools off and the modulation frequency of the laser beam inducing the P_R and P_A powers. In particular, with the laser beam off, that is, when the beam is thermally driven (by the ambient temperature), its response amplitude is random and, therefore, captured by its amplitude spectral density, which shows a peak at its fundamental mechanical resonance frequency (Figure 3.15).

On the other hand, when the modulated laser drives the cantilever beam, the response is seen to be as shown in Figure 3.16. Here, at low frequencies, when the modulation frequency of the laser impinging (heating) the cantilever and causing it to heat up non-uniformly and deflect is so slow that the beam has time to heat up and cool down between pulses, the *photothermal* amplitude is largest; the beam absorbs maximum heat energy, thus attaining maximum deflection, and releases all of the energy, thus reaching its equilibrium state.

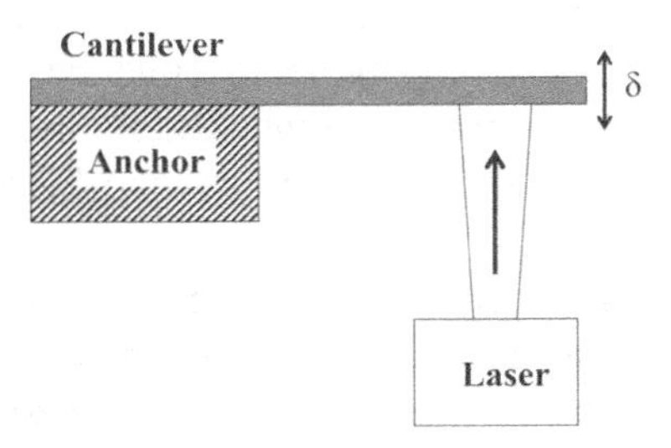

Figure 3.14 Sketch of laser-driven cantilever beam setup.

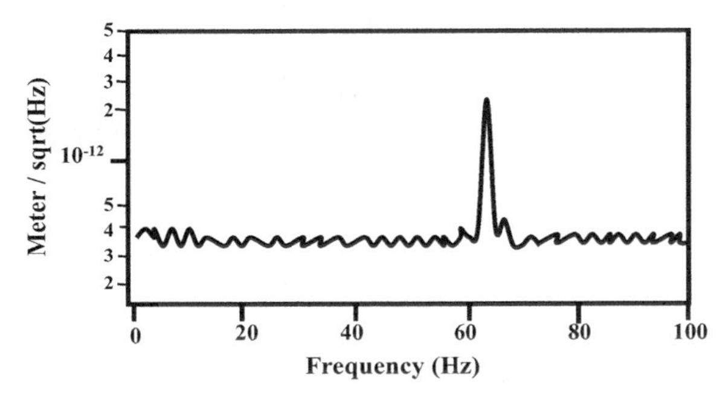

Figure 3.15 Amplitude versus frequency response of thermally driven (laser power is off) of an aluminum-coated silicon cantilever beam in air. The peak occurs at a frequency of 63.6 kHz with quality factor, Q = 155.

Source: Ref. [63].

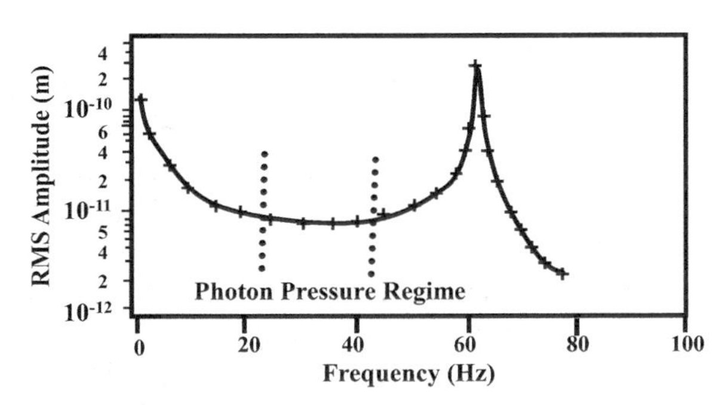

Figure 3.16 Root-mean-square amplitude versus frequency response of laser-driven cantilever beam in air.

Source: Ref. [63].

As the laser modulation frequency increases, a regime is reached in which the beam does not have the time to absorb all the heat energy; thus, its maximum deflection amplitude is less than that at low frequencies, nor to release all the heat energy, so it does not reach equilibrium. This is the "photon pressure regime." Then, for further increases in laser modulation frequency, the beam has no time to absorb heat energy, and its displacement is entirely driven by the radiation pressure, that is, there is no contribution from the laser-induced heating deformation. In this regime, the cantilever beam temperature is constant. At still higher frequencies, the effect of the ambient (thermally driven) mechanical resonance begins to appear, and the modulated laser drive effect is also negligible.

The time-dependent relation between the rate of heating/cooling of the cantilever beam and the modulation rate of the laser beam (Figure 3.16) is clearly a low-pass one, captured by a time constant, τ, so that one can express the frequency dependence of the radiation force as [64],

$$F(\omega) = \frac{F(0)}{(1 + i\omega\tau)} \tag{3.105}$$

and the overall cantilever beam amplitude as,

$$\delta^{rad}(\omega) = \frac{\delta_{rp} + \delta_{pt}(0)/(1 + i\omega\tau)}{1 + \frac{i\omega\tau}{\omega_9 Q} - \left(\frac{\omega}{\omega_0}\right)^2} \tag{3.106}$$

where ω_0 and Q represent the fundamental mechanical resonance frequency and quality factor.

3.2 Mechanical Vibration

Examination of the cantilever beam of Figure 3.4 under distributed electrostatic load conditions reveals its similarity to a springboard. Having the highest load concentration at its tip, one would intuitively expect that the sudden application or removal of the load would lead to mechanical vibration of the beam. Indeed, a mechanical system can, in general, vibrate in n number of modes (geometrical configuration deformations), according to the degrees of freedom it possesses [18].

The degree of freedom of a mechanical system is given by the number of parameters required to specify its position. A springboard, for instance, constrained to move up and down in the z direction has one degree of freedom; a rock sliding across the ice sheet in a curling game has three degrees of freedom, namely the x- and y-displacements of its center of mass and the angle of rotation about its center of gravity; and a basketball en-route to the hoop has six degrees of freedom, namely the x-, y-, and z-displacements of the center of mass and three rotations.

Knowing the number of degrees of freedom of a mechanical system is important because, in analogy with the familiar first- and second-order RL/RC, and RLC circuits of circuit theory [18], they determine how the system responds to an excitation. In particular, normally the goal of the excitation is to couple energy to a particular degree of freedom to produce motion in a particular mode of vibration and not to others. Thus, determining the various vibration modes of a system enables one to modify its architecture so as to diminish or suppress excitation energy coupling to undesirable modes.

In general, mechanical systems may be described by a first-, second-, or nth-order differential equation, depending on whether they are a first-, second-, or nth-degree system, respectively. As it turns out, the higher the number of degrees of freedom of a system, the more complicated it becomes to solve exactly the differential equations describing their motion. Therefore, systems have been classified as few-degrees-of-freedom and many-degrees-of-freedom, with simple exact analysis techniques developed for the former, while specialized approximate techniques are applied to the latter. As a matter of fact, computer-aided numerical techniques implemented in commercial software tools have been developed to analyze general mechanical systems.

Some intuition regarding the relation between structural features, that is, material properties and geometry of a system, may be obtained from simple and approximate analytical analyses of these systems. We next treat techniques for the analysis of a single-degree-of-freedom system and introduce the Rayleigh Quotient technique, developed for analyzing a many-degree-of-freedom system [18].

3.2.1 The Single-Degree-of-Freedom System

For a single-degree-of-freedom mechanical system such as a cantilever beam, the response to an external excitation force is characterized by three parameters, namely its mass, M; its stiffness, denoted by the "spring constant" K; and its damping constant, D. In the most simple case, the response is given by the solution to Newton's second law, $\text{force} = \text{mass} \times \text{acceleration}$. Assuming the displacement in the vertical direction is along the z-axis, and neglecting the gravitational acceleration, if an external z-directed force

$F \sin \omega t$ is applied, the equation of motion of the beam takes the form,

$$M\frac{d^2z}{dt^2} + D\frac{dz}{dt} + Kz = F \sin \omega t \tag{3.107}$$

where t denotes the time. This equation states that the z-directed motion excited by an applied force, $F \sin \omega t$, is the result of the balance between the applied force, and the resultant of the impulsive motion, Md^2z/dt^2, the viscous damping force experienced by the system once it starts to develop a speed, Ddz/dt, and the "spring" force Kz, which attempts to return the system to its equilibrium position. The four terms in Equation (3.107) denote the force of inertia, the damping force, the spring force, and the external force. Equation (3.107) is a second-order linear differential equation with familiar solutions [18]. If the damping can be neglected and the system is unforced, then Equation (3.107) takes the form,

$$M\frac{d^2z}{dt^2} + Kz = 0 \tag{3.108}$$

or

$$\frac{d^2z}{dt^2} = -\frac{K}{M}z \tag{3.109}$$

whose most general solution is,

$$z = A \sin t\sqrt{\frac{K}{M}} + B \cos t\sqrt{\frac{K}{M}} \tag{3.110}$$

where A and B are arbitrary constants. Equation (3.110) embodies the time evolution of the beam's spatial shape under undamped conditions, one cycle of which occurs when $\omega_n = \sqrt{K/M}$, the so-called natural circular frequency varies through $360°$ or 2π radians [18]. The natural angular frequency, f_n, is,

$$f_n = \frac{\omega_n}{2\pi} = \frac{1}{2\pi}\sqrt{\frac{K}{M}} \tag{3.111}$$

and is measured in cycles per second (Hz). The physical interpretation of this frequency is that it captures the fact that, left to itself, a system possessing mass M and spring constant K will vibrate in response to the force exerted by its own mass. Since no external force is required in order to excite these vibrations, they are referred to as free vibrations. This frequency, as will be shown later, is related to the lowest excitation frequency at which the vibration amplitude of the cantilever beam peaks.

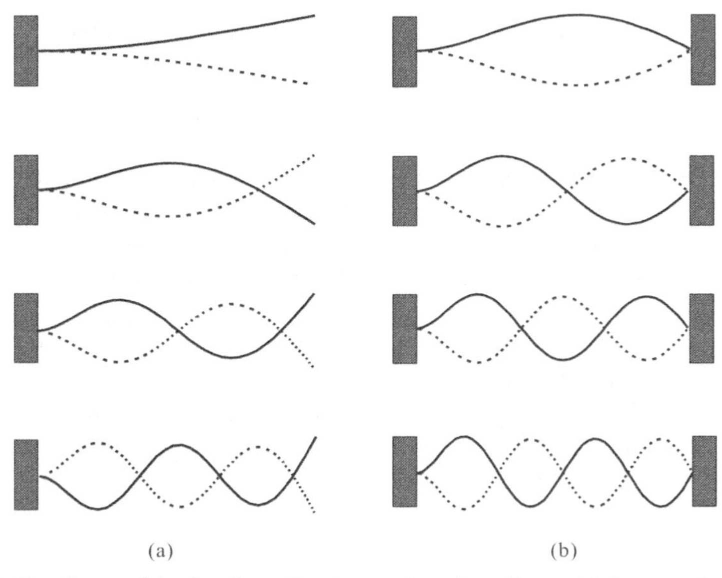

Figure 3.17 Shape of the first four vibration modes of cantilever: (a) Supported on the left and free on the right end. (b) Supported on both left and right ends. Lowest mode shape pictured at the top of the figure.

In general, a mechanical structure has the capability of vibrating at an infinity of frequencies. The sketch of Figure 3.17 depicts the first four modes of vibration of a cantilever beam. In practice, constraints may be introduced to suppress undesired modes.

3.2.2 The Many-Degree-of-Freedom System

Determining the frequencies of the modes at which a many-degree-of-freedom system vibrates, for example, the interdigitated comb-drive capacitor, from the nth-order differential equation describing their motion is usually virtually impossible due to the level of complexity that arises. In such cases, the so-called method of Rayleigh is employed [18].

In the Rayleigh method, instead of solving the equation of motion to determine the configuration adopted by the system at a vibration mode, an equation for the vibration mode is assumed. The vibration frequency then can be calculated in a direct manner from an energy consideration [18]. In particular, as basic mechanics teaches, the kinetic and potential (elastic) energies of a mechanical system oscillate back and forth between them, with the rate at which this energy oscillation occurs being equal to the vibration frequency of the given mode. In this process, in the middle of the motion, when the system passes through its equilibrium configuration, the kinetic energy is maximum and the potential energy is minimum, but when either extreme position is reached, there is zero kinetic energy and maximum potential energy. In this state, the beam is storing all the energy as elastic tension. At any position between the middle and the extreme, both elastic energy and kinetic energy are present, adding in fact to a constant sum if the system is isolated. The vibration or resonance frequency is obtained by equating the kinetic energy in the middle of a vibration to the elastic energy in an extreme position. First, the energies are calculated as follows.

Upon reaching the peak displacement, z_0, the potential or elastic energy due to the spring force is $\int_0^z Kz'dz' = Kz^2/2$. Then, at any instant of time, the kinetic energy is $mv^2/2$. If we assume that the motion is given by $z = z_0 \sin \omega t$, then the corresponding velocity is $v = z_0 \omega \cos \omega t$. On the other hand, the potential energy at the peak displacement is $Kz_0^2/2$, and the kinetic energy in the equilibrium position, where the velocity is maximum, is $1/2mv_{\text{max}}^2 = 1/2m\omega^2 z_0^2$.

Equating energies we have,

$$\frac{1}{2}Kz_0^2 = \frac{1}{2}m\omega^2 z_0^2 \tag{3.112}$$

from which $\omega^2 = K/M$, independent from the amplitude z_0. In this way, the lowest or fundamental frequency is given by the so-called Rayleigh's Quotient,

$$\omega^2 = \frac{1/2Kz_0^2}{1/2mz_0^2} = \frac{V_{\text{max}}}{T_{\text{max}}} \tag{3.113}$$

3.2.3 Rayleigh's Method

Rayleigh's Quotient was derived based on the motion of a beam's tip. The method of Rayleigh generalizes the above procedure to determine the lowest or fundamental frequency of vibration for a general system that possesses distributed mass and flexibility [18]. The governing equation of the structure is given by,

$$EI\frac{\partial^4 u(z,t)}{\partial z^4} = -\mu_1 \frac{\partial^2 u(z,t)}{\partial t^2} \tag{3.114}$$

where E, I, and μ_1 are the modulus of elasticity, the moment of inertia, and the mass per unit length, respectively; $u(z,t)$ is the geometrical deformation of the structure under vibration; and t is the time. The method relies on assuming the shape for the first normal elastic curve for the mechanical deformation corresponding to the maximum displacement of the vibration. We assume that the geometrical deformation of the structure under vibration, $u(z,t)$, is separable into a product of independent spatial and temporal functions, $U(z)$ and $f(t)$, respectively. Then, with the continuum vibration given by [18],

$$u(z,t) = U(z)f(t) \tag{3.115}$$

a continuum velocity given by

$$\dot{u}(z,t) = U(z)\dot{f}(t) \tag{3.116}$$

and a continuum potential given by

$$\frac{\partial u(z,t)}{\partial z} = \frac{\partial U(z)}{\partial z} f(t) \tag{3.117}$$

substituting the above equations into Equation (3.117), we obtain,

$$EI\frac{\partial^3 U(z)}{\partial z^4} = \mu_1 \omega^2 U(z) \tag{3.118}$$

and the corresponding potential energy V and kinetic energy T are given by

$$V = \frac{1}{2}\int_0^L EI\left[\frac{\partial U(z)}{\partial z} f(t)\right]^2 dz \tag{3.119}$$

and

$$T = \frac{1}{2}\int_0^L M(z)\left[U(z)\dot{f}(t)\right]^2 dz \tag{3.120}$$

where $M(z)$ is the distributed mass. The Rayleigh's quotient is then given by,

$$\omega^2 = \frac{\int_0^L EI\left[\frac{\partial U(z)}{\partial z}\right]^2 dz}{\int_0^L M(z)\left[U(z)\right]^2 dz} \tag{3.121}$$

An application of this procedure may be done as follows. We assume the spatial profile of the continuum vibration to be given by the curve,

$$U(z) = z_0\left(1 - \cos\frac{\pi x}{2L}\right) \tag{3.122}$$

Substituting Equation (3.122) into Equation (3.121), we obtain,

$$\omega^2 = \frac{\int_0^L EI \left[\frac{\partial U(z)}{\partial z}\right]^2 dz}{\int_0^L M(z)[U(z)]^2 dz} = \frac{\frac{\pi^4}{64}\frac{EI}{L^3} z_0^2}{\mu_1 z_0^2 L \left(\frac{3}{4} - \frac{2}{\pi}\right)} \tag{3.123}$$

or

$$\omega = \frac{\pi^2}{8\sqrt{\frac{3}{4} - \frac{2}{\pi}}} \sqrt{\frac{EI}{\mu_1 L^4}} = \frac{3.66}{L^2} \sqrt{\frac{EI}{\mu_1}} \tag{3.124}$$

where μ_1 is the mass density per unit length.

Thus, the method of Rayleigh embodies an approximate way for estimating the frequency of the first mode of vibration of a system, based on an assumed motion. Whenever the assumed happens to exact, then the method gives the exact solution for the natural frequency. On the other hand, the assumption of a constant mass, M, and a motion given by $z = z_0 \sin \omega t$ results in a Rayleigh quotient equal to $\omega^2 = K/M$, which is identical to Equation (3.111).

3.3 Thermal Noise in MEMS/NEMS

While the natural frequencies of vibration of a mechanical structure are elicited *spontaneously* by their own weight and give rise to the pristine geometrical shapes characteristic of every mode, in reality there are two other sources that induce vibration. These are [65] (i) intrinsic or dissipative noise, which is driven by *internal fluctuations* due to the structure's temperature; and (ii) extrinsic or non-dissipative noise, which arises when the fluctuations are driven by externally imposed temperature variations such as in the laser-driven cantilever beam previously discussed. Next, following Ref. [66], we expose the fundamental origin of the intrinsic or dissipative noise.

3.3.1 Fundamental Origin of Intrinsic Noise [66]

As is well known, microscopic particles immersed in a fluid held at a finite temperature will exhibit a random motion, denoted as *Brownian motion*, which is due to collisions with the randomly moving atoms and molecules making up the fluid. In particular, since the motion of the atoms and molecules obeys kinetic theory, the average translational energy of each particle equals $\frac{3}{2}k_B T$, in accord with the equipartition of energy [67].

Similarly, the atoms in a solid (e.g., a cantilever beam held at a finite temperature) experience a random motion [46]. This random motion has a certain frequency distribution which, thus, alters the fundamental vibration frequencies derived in the previous sections. In addition, in applications where the cantilever beam is part of, for example, a tuning capacitor, the random motion will manifest as a random variation in the capacitance which, in turn, will manifest as a random variation in the frequency it is attempting to set up. The question, then, is: What is the amplitude and frequency spectrum of this random vibration due to a finite temperature and how is it related to the beam properties? The answer to that question, encapsulated in the so-called *fluctuation-dissipation theorem*, was derived in 1951 by Callen and Welton and will be discussed subsequently.

We begin by stating the following preliminaries:

(1) A system in a quantum state with wave function Ψ_n has an energy E_n.
(2) The probability that a system at a temperature T occupies an energy E_n is given by the Boltzmann factor,

$$P_n = e^{-\frac{E_n}{k_B T}} \tag{3.125}$$

(3) The probabilities of occupancy of the energy levels $E_n, E_n + \hbar\omega$, and $E_n - \hbar\omega$ are related by the weighting factor $f(E)$ such that,

$$\frac{f(E_n + \hbar\omega)}{f(E_n)} = \frac{f(E_n)}{f(E_n - \hbar\omega)} = e^{-\frac{\hbar\omega}{k_B T}} \tag{3.126}$$

(4) The density of states as a function of energy is denoted by $\rho(E)$ such that the number of quantum states between the energy E and the energy $E + \delta E$, in the neighborhood of E, is $\rho(E)\delta E$.
(5) A system at an energy state E_n may effect a transition to an energy level $E_n + \hbar\omega$ if it *absorbs* an energy $\hbar\omega$, and may effect a transition to an energy level $E_n - \hbar\omega$ if it *emits* an energy $\hbar\omega$.

At any given time, the occupancy of the energy level E_n is a result of the *net transition rate.*

Now, if the system is unperturbed, the *stationary* wave functions Ψ_n are the solution to the Schrödinger equation,

$$H_0\Psi_n = E_n\Psi_n \tag{3.127}$$

where H_0, assumed to be a function of position $q_1, \ldots q_k \ldots$ and momenta $p_1, \ldots, p_k \ldots$, is the unperturbed Hamiltonian. However, when the system is perturbed by, say, a disturbance $Q(q_1, \ldots q_k \ldots, p_1, \ldots, p_k \ldots)$, then its Hamiltonian is given by,

$$H = H_0(\ldots p_k \ldots q_k \ldots) + VQ(\ldots p_k \ldots q_k \ldots) \tag{3.128}$$

where V, which may be expressed by,

$$V = V_0 \sin \omega t \tag{3.129}$$

is a function of time that captures the instantaneous magnitude of the perturbation. Under a perturbation, the wave function of the system $\Psi(t)$ is given as a linear combination of the stationary wave functions Ψ_n [68],

$$\Psi(t) = \sum_n a_n(t)\Psi_n = \sum_n a_n(t)\Psi_n^0(q)e^{-\frac{E_n t}{\hbar}} \tag{3.130}$$

where the coefficients $a_n(t)$ are obtained by integrating the time-dependent Schrödinger equation,

$$H\Psi = i\hbar\frac{\partial \Psi}{\partial t} \tag{3.131}$$

or

$$H_0\Psi + V_0 \sin \omega t Q\Psi = i\hbar\frac{\partial \Psi}{\partial t} \tag{3.132}$$

Suppose we express Equation (3.132) as,

$$[H_0 + U_S]\Psi = i\hbar\frac{\partial \Psi}{\partial t} \tag{3.133}$$

with

$$U_S(q,t) = V_0 \sin \omega t Q = V_0 Q\left[\frac{e^{i\omega t} - e^{-i\omega t}}{2i}\right] \tag{3.134}$$

embodying the perturbing "scattering" potential U_S that causes the system to transition from the state n to the state n'. Then, assuming that at the initial time, $t = 0$, the system is in state n, so that $a_n(t = 0) = 1$, and the state n' is unoccupied, so that $a_{n'}(t = 0) = 0$, the probability of the $n \rightarrow n'$ transition is given by,

$$Pn \rightarrow n' = \lim_{t\rightarrow\infty} |a_{n'}(t)|^2 \tag{3.135}$$

and the transition rate is given by,

$$Rn \rightarrow n' = \frac{\lim_{t\rightarrow\infty} |a_{n'}(t)|^2}{t} \tag{3.136}$$

Now, to find $a_n(i)$ one substitutes Equation (3.130) into Equation (3.133), which results in,

$$U_S \sum_n a_n(t) \Psi_n^0 e^{-i\frac{E_n t}{\hbar}} = i\hbar \sum_n \frac{\partial a_n}{\partial t} \Psi_n^0 e^{-i\frac{E_n t}{\hbar}} \tag{3.137}$$

Then, multiplying by $\Psi_{n'}^{0*} e^{i\frac{E_{n'} t}{\hbar}}$, integrating, and making use of orthogonality of the wave functions, one obtains,

$$i\hbar \frac{\partial a_{n'}}{\partial t} = \sum_n H_{n'n} a_n(t) e^{i\frac{[E_{n'} - E_n]t}{\hbar}} \tag{3.138}$$

where

$$H_{n'n}(t) = \int_{-\infty}^{+\infty} \Psi_{n'}^{0*}(q) U_S(q,t) \Psi_n^0(q) dq \tag{3.139}$$

is the matrix element of the perturbation potential that causes the system to transition from state n to state n'.

To proceed, Equation (3.138) may be reduced to one term if one assumes that $a_n(t = 0) = 1$ and the occupancy of all other states is negligible. Then, we have the so-called Born approximation,

$$i\hbar \frac{\partial a_{n'}}{\partial t} = H_{n',n} a_n(t) e^{i\frac{[E_{n'} - E_n]t}{\hbar}} \tag{3.140}$$

which can be solved by integrating, to obtain,

$$a_{n'}(t) = \frac{1}{i\hbar} \int_0^t H_{n'n}(t) e^{i\frac{[E_{n'} - E_n]t'}{\hbar}} dt' \tag{3.141}$$

where

$$\begin{aligned} H_{n'n}(t) &= \int_{-\infty}^{+\infty} \Psi_{n'}^{0*}(q) V_0 Q(q) \left[\frac{e^{i\omega t} - e^{i\omega t}}{2i} \right] \Psi_n^0(q) dq \\ &= \frac{V_0}{2i} \int_{-\infty}^{+\infty} \Psi_{n'}^{0*}(q) Q(q) e^{i\omega t} \Psi_n^0(q) dq \\ &\quad - \frac{V_0}{2i} \int_{-\infty}^{+\infty} \Psi_{n'}^{0*}(q) Q(q) e^{-i\omega t} \Psi_n^0(q) dq \\ &= H_{n'n} e^{i\omega t} - H_{n'n} e^{-i\omega t} \end{aligned} \tag{3.142}$$

Now, substituting Equation (3.142) back into Equation (3.141), we obtain,

$$
\begin{aligned}
a_{n'}(t) &= \frac{1}{i\hbar}\int_0^t H_{n'n}(t)e^{i\frac{[E_{n'}-E_n]t'}{\hbar}}dt' \\
&= \frac{1}{i\hbar}\int_0^t [H_{n'n}e^{i\omega t} - H_{n'n}e^{-i\omega t}]e^{i\frac{[E_{n'}-E_n]t'}{\hbar}}dt' \\
&= \frac{1}{i\hbar}\int_0^t \left[H_{n'n}e^{i\frac{[E_{n'}-E_n+\hbar\omega]t'}{\hbar}} - H_{n'n}e^{i\frac{[E_{n'}-E_n-\hbar\omega]t'}{\hbar}}\right]dt' \\
&= \frac{1}{i\hbar}H_{n'n}\frac{e^{i\frac{[E_{n'}-E_n+\hbar\omega]t}{\hbar}}-1}{i\frac{[E_{n'}-E_n+\hbar\omega]}{\hbar}} - \frac{1}{i\hbar}H_{n'n}\frac{e^{i\frac{[E_{n'}-E_n-\hbar\omega]t}{\hbar}}-1}{i\frac{[E_{n'}-E_n-\hbar\omega]}{\hbar}}
\end{aligned}
\tag{3.143}
$$

which, defining,

$$\Delta^- = E_{n'} - (E_n - \hbar\omega) \tag{3.144}$$

and

$$\Delta^+ = E_{n'} - (E_n + \hbar\omega) \tag{3.145}$$

allows us to express,

$$a_{n'}(t) = \frac{1}{i\hbar}H_{n'n}\frac{e^{i\frac{\Delta^- t}{\hbar}}-1}{\frac{i\Delta^-}{\hbar}} - \frac{1}{i\hbar}H_{n'n}\frac{e^{i\frac{\Delta^+ t}{\hbar}}-1}{\frac{i\Delta^-}{\hbar}} \tag{3.146}$$

as

$$
\begin{aligned}
a_{n'}(t) &= \frac{1}{i\hbar}H_{n'n}\frac{e^{i\frac{\Delta^- t}{\hbar}}-1}{\frac{i\Delta^-}{\hbar}} - \frac{1}{i\hbar}H_{n'n}\frac{e^{i\frac{\Delta^+ t}{\hbar}}-1}{\frac{i\Delta^+}{\hbar}} \\
&= \frac{1}{i\hbar}H_{n'n}e^{-\frac{i\Delta^- t}{2\hbar}}2\frac{e^{\frac{i\Delta^- t}{2\hbar}} - e^{-\frac{i\Delta^- t}{2\hbar}}}{i2\frac{\Delta^-}{\hbar}} \\
&\quad - \frac{1}{i\hbar}H_{n'n}e^{-\frac{i\Delta^+ t}{2\hbar}}2\frac{e^{\frac{i\Delta^+ t}{2\hbar}} - e^{-\frac{i\Delta^+ t}{2\hbar}}}{i2\frac{\Delta^+}{\hbar}}
\end{aligned}
\tag{3.147}
$$

$$
\begin{aligned}
a_{n'}(t) &= \frac{1}{i\hbar}H_{n'n}e^{-\frac{i\Delta^- t}{2\hbar}}\frac{\sin\left(\frac{\Delta^- t}{2\hbar}\right)}{\frac{\Delta^- t}{2\hbar}}t \\
&\quad - \frac{1}{i\hbar}H_{n'n}e^{-\frac{i\Delta^+ t}{2\hbar}}\frac{\sin\left(\frac{\Delta^+ t}{2\hbar}\right)}{\frac{\Delta^+ t}{2\hbar}}t
\end{aligned}
$$

Then, the transition rate,

$$R^{-} = \frac{\lim_{t\to\infty} |a_{n'}^{-}(t)|^2}{t} = \frac{\lim_{t\to\infty} \left| \frac{1}{i\hbar} H_{n'n} e^{-\frac{i\Delta^{-}t}{2\hbar}} \frac{\sin\left(\frac{\Delta^{-}t}{2\hbar}\right)}{\frac{\Delta^{-}t}{2\hbar}} t \right|^2}{t}$$

$$= \frac{\lim_{t\to\infty} |H_{n'n}|^2 \left[\frac{\sin\left(\frac{\Delta^{-}t}{2\hbar}\right)}{\frac{\Delta^{-}t}{2\hbar}} \right]^2 t^2}{\hbar^2 t} \tag{3.148}$$

Since the $(\sin(x)/x)^2$ or $sinc(x)$ function is tall and narrow, it may be used to represent the Dirac delta function, that is,

$$\delta(\Delta) = \left(\frac{\sin\left(\frac{\Delta t}{2\hbar}\right)}{\frac{\Delta t}{2\hbar}} \right)^2 \tag{3.149}$$

if the area of the expression on the left-hand side of the equality sign is equal to the area of the expression on the right-hand side. This will occur if,

$$\int_{-\infty}^{+\infty} \delta(\Delta) d\Delta = \int_{-\infty}^{+\infty} \left[\frac{\sin\left(\frac{\Delta t}{2\hbar}\right)}{\frac{\Delta t}{2\hbar}} \right]^2 d\Delta \tag{3.150}$$

Transforming variables on the right we have,

$$\alpha = \frac{\Delta t}{2\hbar} \to \frac{2\hbar}{t}\alpha = \Delta \to \frac{2\hbar}{t} d\alpha = d\Delta \tag{3.151}$$

so, the integral becomes,

$$\int_{-\infty}^{+\infty} \delta(\Delta) d\Delta = \frac{2\hbar}{t} \int_{-\infty}^{+\infty} \left[\frac{\sin(\alpha)}{\alpha} \right]^2 d\alpha = \frac{2\hbar\pi}{t} \tag{3.152}$$

since the integral of the squared $sinc(\alpha)$ function is π. So, we have,

$$\int_{-\infty}^{+\infty} \left[\frac{\sin\left(\frac{\Delta t}{2}\right)}{\frac{\Delta t}{2}} \right]^2 d\Delta = \frac{2\pi\hbar}{t} \tag{3.153}$$

Thus, replacing the squared $sinc$ function by,

$$\left(\frac{\sin\left(\frac{\Delta t}{2\hbar}\right)}{\frac{\Delta t}{2\hbar}}\right)^2 = \frac{2\hbar\pi}{t}\delta(\Delta) \tag{3.154}$$

we have,

$$R^- = \frac{|H_{n'n}|^2 \frac{2\hbar\pi}{t}\delta(\Delta)t^2}{\hbar^2 t} = \frac{2\pi}{\hbar}|H_{n'n}|^2\delta(E_{n'} - (E_n - \hbar\omega)) \tag{3.155}$$

and similarly,

$$R^+ = \frac{2\pi}{\hbar}|H_{n'n}|^2\delta(E_{n'} - (E_n + \hbar\omega)) \tag{3.156}$$

Now, with,

$$H_{n'n} = \frac{V_0}{2i}\int_{-\infty}^{+\infty} \Psi_{n'}^{0*}(q)Q(q)\Psi_n^0(q)dq = \frac{V_0}{2i}\overline{Q} \tag{3.157}$$

we have,

$$R^- = \frac{2\pi}{\hbar}\left|\frac{V_0}{2i}\overline{Q}\right|^2 \delta(E_{n'} - (E_n - \hbar\omega)) = \frac{\pi V_0^2}{2\hbar}\left|\overline{Q}\right|^2 \delta(E_{n'} - (E_n - \hbar\omega)) \tag{3.158}$$

This is the rate at which the system transitions from an energy state E_n to a lower energy state $E_{n'} = E_n - \hbar\omega$ by emitting an energy $\hbar\omega$. If the energy states are infinitesimally close,

$$R^- = \frac{\pi V_0^2}{2\hbar}\left|\langle E_n - \hbar\omega\left|\overline{Q}\right| E_n\rangle\right|^2 \rho(E_n - \hbar\omega) \tag{3.159}$$

and, similarly, for the rate of absorption of an energy $\hbar\omega$, we have,

$$R^+ = \frac{\pi V_0^2}{2\hbar}\left|\langle E_n + \hbar\omega\left|\overline{Q}\right| E_n\rangle\right|^2 \rho(E_n + \hbar\omega) \tag{3.160}$$

Since the system's energy increases or decreases depending on whether the energy $\hbar\omega$ is absorbed or emitted, respectively, the net energy absorbed by the system per unit time, that is, the power dissipated by it is,

$$\begin{aligned} P'_{net} = \hbar\omega(R^+ - R^-) = \frac{\pi V_0^2\omega}{2}\Big(&\left|\langle E_n + \hbar\omega\left|\overline{Q}\right| E_n\rangle\right|^2 \rho(E_n + \hbar\omega) \\ &- \left|\langle E_n - \hbar\omega\left|\overline{Q}\right| E_n\rangle\right|^2 \rho(E_n - \hbar\omega)\Big) \end{aligned} \tag{3.161}$$

In a real system, the likelihood that a state is occupied diminishes at higher energies. To take this into account, the above equation is multiplied by $f(E)$ to give the actual power dissipated,

$$\overline{P} = \frac{\pi V_0^2 \omega}{2} \sum_n \Big(\left| \langle E_n + \hbar\omega \left| \overline{Q} \right| E_n \rangle \right|^2 \rho(E_n + \hbar\omega) - \left| \langle E_n - \hbar\omega \left| \overline{Q} \right| E_n \rangle \right|^2 \rho(E_n - \hbar\omega) \Big) \cdot f(E_n) \tag{3.162}$$

As usual [68], the evaluation of summation over n may be replaced with integration. Doing so, we have,

$$\overline{P} = \frac{\pi V_0^2 \omega}{2} \int_0^{\infty} \Big(\left| \langle E + \hbar\omega \left| \overline{Q} \right| E \rangle \right|^2 \rho(E + \hbar\omega) - \left| \langle E - \hbar\omega \left| \overline{Q} \right| E \rangle \right|^2 \rho(E - \hbar\omega) \Big) \cdot \rho(E) f(E) dE \tag{3.163}$$

In this way, Callen and Welton showed that the application of a periodic perturbation to a system results in a dissipation that is quadratic in the perturbation.

Now, to relate the dissipation to a property of the system, one posits the applicability of the well-known fact that if the perturbation is small, the response is proportional to a linear function of it. In particular, the response $\dot{\overline{Q}}$ (current) to an applied force V (voltage) may be expressed by,

$$V = Z(\omega)\dot{\overline{Q}} \tag{3.164}$$

that is, Ohm's law. This may be related to the instantaneously dissipated power using,

$$P = V\dot{\overline{Q}} \cdot \frac{R}{|Z|} \tag{3.165}$$

that is, the real instantaneous power dissipated is the product of the voltage and current with the fraction of the impedance that is resistive $R/|Z|$. On the other hand, the average power dissipated is given by,

$$\overline{P} = \mathrm{Re}\left\{ \frac{1}{T} \int_0^T V \cdot \dot{\overline{Q}} dt \right\} = \mathrm{Re}\left\{ \frac{1}{T} \int_0^T V_0 \sin\omega t \cdot \frac{V_0 \sin\omega t}{Z(\omega)} dt \right\}$$

$$= \mathrm{Re}\left\{\frac{V_0^2}{TZ(\omega)}\int_0^T \sin^2\omega t dt\right\} = \frac{V_0^2}{T}\int_0^T \sin^2\omega t dt \cdot \mathrm{Re}\left\{\frac{1}{Z(\omega)}\right\}$$
$$= \frac{V_0^2}{T}\cdot \mathrm{Re}\left\{\frac{1}{Z(\omega)}\cdot\frac{Z^*(\omega)}{Z^*(\omega)}\right\} = \frac{V_0^2}{2|Z(\omega)|^2}\cdot \mathrm{Re}\{Z^*(\omega)\}$$
$$= \frac{V_0^2}{T}\cdot\frac{R}{|Z(\omega)|^2} \tag{3.166}$$

Equating Equation (3.166) with Equation (3.163) we obtain,

$$\frac{V_0^2}{2}\cdot\frac{R}{|Z|^2} = \frac{\pi V_0^2\omega}{2}\int_0^\infty \Big(\left|\langle E+\hbar\omega\left|\overline{Q}\right| E\rangle\right|^2\rho(E+\hbar\omega) - \left|\langle E-\hbar\omega\left|\overline{Q}\right| E\rangle\right|^2\rho(E-\hbar\omega)\Big)\cdot\rho(E)f(E)dE \tag{3.167}$$

and, consequently,

$$\frac{R}{|Z|^2} = \pi\omega\int_0^\infty \Big(\left|\langle E+\hbar\omega\left|\overline{Q}\right| E\rangle\right|^2\rho(E_n+\hbar\omega) - \left|\langle E-\hbar\omega\left|\overline{Q}\right| E\rangle\right|^2\rho(E-\hbar\omega)\Big)\cdot\rho(E)f(E)dE \tag{3.168}$$

Next, we expose the fluctuation part of the theorem.

In this case, instead of the system experiencing a response $\dot{\overline{Q}}$ as a result of the applied force, V, it is assumed that the system is not subject to any externally applied force, but that despite being in equilibrium, it experiences a *spontaneously fluctuating force* that causes $\langle\dot{\overline{Q}}\rangle = 0$, but $\langle\dot{\overline{Q}}^2\rangle \neq 0$; the consequence of this is now determined.

To find $\langle\dot{\overline{Q}}^2\rangle$, the mean square fluctuation of the disturbance Q, one assumes that it is elicited by a spontaneously fluctuating force, V, characterized by $\langle V^2\rangle$, its corresponding mean squared fluctuating force. From the relationship,

$$V = Z(\omega)\dot{\overline{Q}} \tag{3.169}$$

we postulate that,

$$\langle V^2\rangle = Z(\omega)\langle\dot{\overline{Q}}^2\rangle \tag{3.170}$$

from which we obtain $\langle V^2\rangle$. Now, we obtain $\langle \overline{\dot{Q}}^2\rangle$, from the definition of mean square fluctuation,

$$\langle(\overline{\dot{Q}} - \langle\overline{\dot{Q}}\rangle)^2\rangle = \langle(\overline{\dot{Q}}^2 - 2\overline{\dot{Q}}\langle\overline{\dot{Q}}\rangle + \langle\overline{\dot{Q}}\rangle^2)^2\rangle = \langle\overline{\dot{Q}}^2\rangle \tag{3.171}$$

given that we assumed $\langle\overline{\dot{Q}}\rangle = \langle E_n|\overline{\dot{Q}}|E_n\rangle = 0$ for a state E_n. In turn, for the mean square fluctuation we have,

$$\langle\overline{\dot{Q}}^2\rangle = \sum_m \langle E_n|\dot{Q}|E_m\rangle\langle E_m|\dot{Q}|E_n\rangle \tag{3.172}$$

Now, from the equation for the time derivative of a quantum operator,

$$\langle\dot{Q}\rangle = \frac{d\langle Q\rangle}{dt} = \frac{1}{i\hbar}\langle[H, Q]\rangle \tag{3.173}$$

we have,

$$\begin{aligned}\langle\overline{\dot{Q}}^2\rangle &= -\frac{1}{\hbar^2}\sum_m \langle E_n|H_0Q - QH_0|E_m\rangle \times \langle E_m|H_0Q - QH_0|E_n\rangle \\ &= \hbar^{-2}\sum_m (E_n - E_m)^2|\langle E_n|Q|E_m\rangle|^2 \end{aligned}\tag{3.174}$$

Next, denoting,

$$\hbar\omega = |E_n - E_m| \tag{3.175}$$

the summation over m is replaced by integrals for $E_n < E_m$ and $E_n > E_m$ to have,

$$\begin{aligned}\langle E_n|\overline{\dot{Q}}^2|E_m\rangle &= \hbar^{-2}\int_0^\infty (\hbar\omega)^2|\langle E_n + \hbar\omega|Q|E_n\rangle|^2\rho(E_n + \hbar\omega)\hbar d\omega \\ &\quad + \hbar^{-2}\int_0^\infty (\hbar\omega)^2|\langle E_n - \hbar\omega|Q|E_n\rangle|^2\rho(E_n - \hbar\omega)\hbar d\omega \\ &= \hbar\omega^2\left\{\int_0^\infty \langle E_n + \hbar\omega|Q|E_n\rangle|^2\rho(E_n + \hbar\omega)\right. \\ &\quad \left. + |\langle E_n - \hbar\omega|Q|E_n\rangle|^2\rho(E_n - \hbar\omega)\right\} d\omega \end{aligned}\tag{3.176}$$

In a real thermodynamic system, the fluctuation observed must include the impact of the occupation factor, $f(E_n)$, by way of weighting each level E_n

and summing over all E_n levels. This is expressed as,

$$\langle|\dot{Q}^2|\rangle = \sum_n f(E_n)\left\{\hbar\omega^2 \int_0^\infty |\langle E_n + \hbar\omega|Q|E_n\rangle|^2 \rho(E_n + \hbar\omega) + |\langle E_n - \hbar\omega|Q|E_n\rangle|^2 \rho(E_n - \hbar\omega)\right\} d\omega \tag{3.177}$$

As previously done, we can replace summation by integration by multiplying the integrand by the density of states and integrating; this gives,

$$\langle|\dot{Q}^2|\rangle = \int_0^\infty \hbar\omega^2 \int_0^\infty [\rho(E)f(E)\{|\langle E + \hbar\omega|Q|E\rangle|^2 \rho(E + \hbar\omega) + |\langle E - \hbar\omega|Q|E\rangle|^2 \rho(E - \hbar\omega)\}dE]d\omega \tag{3.178}$$

Finally, utilizing the impedance definition, we have,

$$\langle V^2\rangle = \int_0^\infty |Z|^2\hbar\omega^2 \int_0^\infty [\rho(E)f(E)\{|\langle E + \hbar\omega|Q|E\rangle|^2 \rho(E + \hbar\omega) + |\langle E - \hbar\omega|Q|E\rangle|^2 \rho(E - \hbar\omega)\}dE] \tag{3.179}$$

Thus far, expressions for $R/|Z|^2$ and $\langle V^2\rangle$, embodying dissipation and fluctuation, respectively, have been obtained. Observation of the respective expressions indicates that they contain the following formulas in common,

$$\int_0^\infty [\rho(E)f(E)\{|\langle E + \hbar\omega|Q|E\rangle|^2 \rho(E + \hbar\omega) \pm |\langle E - \hbar\omega|Q|E\rangle|^2 \rho(E - \hbar\omega)\}dE] \tag{3.180}$$

where the negative sign is associated with $R/|z|^2$ and the positive sign with $\langle V^2\rangle$. Therefore, a relation between dissipation, by way of $R(\omega)$, and fluctuation, by way of $\langle V^2\rangle$, is within reach. To make the relation more explicit, denoting the negative expression by,

$$C(-) = \int_0^\infty \left[\rho(E)f(E)\left\{|\langle E + \hbar\omega|Q|E\rangle|^2 \rho(E + \hbar\omega) - |\langle E - \hbar\omega|Q|E\rangle|^2 \rho(E - \hbar\omega)\right\} dE\right] \tag{3.181}$$

and noticing that in the second integral $|\langle E - \hbar\omega|Q|E\rangle| = 0$ for $E < \hbar\omega$, when the transformation $E \to E + \hbar\omega$ is made in the integration variable, one gets,

$$
\begin{aligned}
C(-) = & \int_0^\infty [\rho(E)f(E)\{|\langle E+\hbar\omega|Q|E\rangle|^2\rho(E+\hbar\omega) \\
& - |\langle E|Q|E+\hbar\omega\rangle|^2\rho(E+\hbar\omega-\hbar\omega)\}dE] \\
= & \int_0^\infty [\rho(E)f(E)\{|\langle E+\hbar\omega|Q|E\rangle|^2\rho(E+\hbar\omega) \\
& - |\langle E|Q|E+\hbar\omega\rangle|^2\rho(E)\}dE] \\
= & \int_0^\infty [\rho(E)f(E)\{|\langle E+\hbar\omega|Q|E\rangle|^2\{\rho(E+\hbar\omega) - \rho(E)\}dE] \\
= & \int_0^\infty \Big[\rho(E)f(E)|\langle E+\hbar\omega|Q|E\rangle|^2 \\
& \rho(E+\hbar\omega)\left\{1 - \frac{\rho(E)}{\rho(E+\hbar\omega)}\right\}dE\Big] \\
= & \int_0^\infty \Big[|\langle E+\hbar\omega|Q|E\rangle|^2 \\
& \rho(E+\hbar\omega)\rho(E)f(E)\left\{1 - \frac{\rho(E)}{\rho(E+\hbar\omega)}\right\}dE\Big]
\end{aligned}
\tag{3.182}
$$

Examination of the term, $\rho(E)/\rho(E+\hbar\omega)$, leads to,

$$
\begin{aligned}
\frac{\rho(E)}{\rho(E+\hbar\omega)} & \Rightarrow \frac{\rho(E)}{\rho(E+\hbar\omega)} \cdot \frac{f(E)}{f(E)}|_E \to E - \hbar\omega \\
& \Rightarrow \frac{\rho(E-\hbar\omega)}{\rho(E-\hbar\omega+\hbar\omega)} \cdot \frac{f(E-\hbar\omega)}{f(E-\hbar\omega)} \\
& \Rightarrow \frac{\rho(E-\hbar\omega)}{\rho(E)} \cdot \frac{f(E-\hbar\omega)}{f(E)} \cdot e^{-\frac{\hbar\omega}{kT}} \cong e^{-\frac{\hbar\omega}{kT}}
\end{aligned}
\tag{3.183}
$$

where we used,

$$
\frac{f(E)}{f(E-\hbar\omega)} = e^{-\frac{\hbar\omega}{k_B T}} \tag{3.184}
$$

and thus we can write,

$$C(-) = \int_0^\infty \Big[|\langle E + \hbar\omega|Q|E\rangle|^2 \rho(E+\hbar\omega)\rho(E)f(E)\left\{1 - e^{-\frac{\hbar\omega}{k_B T}}\right\} dE \Big] \tag{3.185}$$

or

$$C(-) = \left\{1 - e^{-\frac{\hbar\omega}{k_B T}}\right\} \int_0^\infty [|\langle E + \hbar\omega|Q|E\rangle|^2 \rho(E+\hbar\omega)\rho(E)f(E)dE] \tag{3.186}$$

Employing similar arguments for the positive sign one obtains,

$$C(+) = \left(1 + e^{-\frac{\hbar\omega}{kT}}\right) \int_0^\infty [\{|\langle E + \hbar\omega|Q|E\rangle|^2 \rho(E+\hbar\omega)\rho(E)f(E)\}dE] \tag{3.187}$$

Now, reinserting these expressions into those for $R/|Z|^2$ and $\langle V^2\rangle$, one obtains,

$$\frac{R}{|Z|^2} = \pi\omega\left(1 - e^{-\frac{\hbar\omega}{kT}}\right) \times \int_0^\infty [\{|\langle E + \hbar\omega|Q|E\rangle|^2 \rho(E+\hbar\omega)\rho(E)f(E)\}dE] \tag{3.188}$$

from where it follows that,

$$|Z|^2 = \frac{R}{\pi\omega\left(1 - e^{-\frac{\hbar\omega}{kT}}\right) \times \int_0^\infty [\{|\langle E + \hbar\omega|Q|E\rangle|^2 \rho(E+\hbar\omega)\rho(E)f(E)\}dE]} \tag{3.189}$$

and

$$\langle V^2\rangle = \int_0^\infty \frac{R \times \hbar\omega^2\left(1 + e^{-\frac{\hbar\omega}{kT}}\right) \times \int_0^\infty [\{|\langle E + \hbar\omega|Q|E\rangle|^2 \rho(E+\hbar\omega)\rho(E)f(E)\}dE]d\omega}{\pi\omega\left(1 - e^{-\frac{\hbar\omega}{kT}}\right) \times \int_0^\infty [\{|\langle E + \hbar\omega|Q|E\rangle|^2 \rho(E+\hbar\omega)\rho(E)f(E)\}dE]} \tag{3.190}$$

or

$$\langle V^2\rangle = \int_0^\infty \frac{R(\omega) \times \hbar\omega^2\left(1 + e^{-\frac{\hbar\omega}{kT}}\right)d\omega}{\pi\omega\left(1 - e^{-\frac{\hbar\omega}{kT}}\right)} = \int_0^\infty \frac{R \times \hbar\omega\left(1 + e^{-\frac{\hbar\omega}{kT}}\right)d\omega}{\pi\left(1 - e^{-\frac{\hbar\omega}{kT}}\right)} \tag{3.191}$$

But,

$$\frac{1+e^{-\frac{\hbar\omega}{kT}}}{1-e^{-\frac{\hbar\omega}{kT}}} = 2\left(\frac{1}{2} + \frac{1}{e^{\frac{\hbar\omega}{kT}} - 1}\right) \tag{3.192}$$

so, we have

$$\langle V^2 \rangle = \frac{2}{\pi}\int_0^\infty R(\omega)E(\omega,T)d\omega \tag{3.193}$$

where,

$$E(\omega,T) = \frac{1}{2}\hbar\omega + \frac{\hbar\omega}{e^{\frac{\hbar\omega}{kT}} - 1} \tag{3.194}$$

The expression for $E(\omega, T)$ is identified as the mean energy of an oscillator of frequency ω at a temperature T; this is the frequency distribution we were after. If the temperature is high, for example, $\hbar\omega \ll kT$, then,

$$E(\omega,T) \approx kT \tag{3.195}$$

which implies that, at a given temperature, the frequency spectrum is constant or *white*. The average of the square of the voltage fluctuation is then given by,

$$\langle V^2 \rangle = \frac{2}{\pi}kT\int_0^\infty R(\omega)d\omega \tag{3.196}$$

which is the more familiar Nyquist relationship utilized in electronics noise calculations.

The key to relating fluctuation to dissipation is that the Boltzmann factor $\exp(-\hbar\omega/kT)$ allows the calculation of the absorption and emission integrals to be expressed by two factors, namely one that captures absorption, $1 - \exp(-\hbar\omega/kT)$, or emission, $1 + \exp(-\hbar\omega/kT)$, and thus different for the two cases, and one factor that is identical in the two cases, which cancels when relating $R/|Z|^2$ to $\langle V^2 \rangle$. This is not obvious or intuitive!

3.3.1.1 Amplitude of brownian (random) displacement of cantilever beam [69]

Going back to the beginning of the previous section, we recall that we are after the intrinsic or thermal noise of a beam as manifested in its random displacement at a given temperature, which was found to be related to the Brownian noise. But the Brownian noise is the fundamental theory that captures the motion of a small particle immersed in a fluid due to a randomly fluctuating force, $F(t)$, with components F_x, F_y, and F_z, such that in, say, the x-direction, we have,

$$\langle F_x^2 \rangle = \frac{2}{\pi} kT\eta \int_0^{\omega_{Highest}} d\omega \quad (3.197)$$

where η is the friction (analogous to the resistance R in Equation (3.194) defining the frictional force,

$$Frictional\ \ Force = -\eta v \quad (3.198)$$

where v is the particle velocity.

The average of the square of the random fluctuating force at a given temperature, Equation (3.197), is proportional to all the frequencies exhibited in the motion of the particle. In the case of a beam, this implies the frequencies of all its vibration modes. Next, we present the random amplitude calculation of a beam derived by Butt and Jaschke [69] for the two types of beams usually encountered in applications, namely the singly anchored (cantilever) one (Figure 3.18) and the doubly anchored one.

The singly anchored beam is characterized by a length, width, and thickness L, W, and h, respectively, and is made up of a material with a modulus of elasticity E, as a result of which it exhibits a spring constant, $K = 0.25EWh^3/L^3$. K, as is known by now, enters into the determination of the beam vibration amplitudes and frequencies, both being derived by solving the differential equation describing the transversal displacement of the beam, neglecting damping effects [69],

$$\frac{d^2z}{dt^2} + \frac{Eh^2}{12\rho}\frac{d^4z}{dx^4} = 0 \quad (3.199)$$

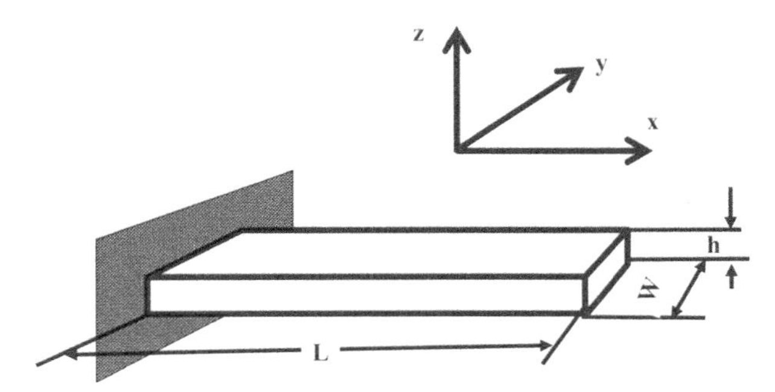

Figure 3.18 Singly anchored beam.

where ρ is the beam density. The general solution to this equation is the beam transversal amplitude vibration, z, given by [69],

$$z = \sum_{i=1}^{\infty} C_i \sin(\omega_i t + \delta_i) \cdot \Phi_i \tag{3.200}$$

where

$$\begin{aligned} \Phi_i = &(\sin \alpha_i + \sinh \alpha_i) \left(\cos \frac{\alpha_i}{L} x - \cosh \frac{\alpha_i}{L} x \right) \\ &- (\cos \alpha_i + \cosh \alpha_i) \left(\sin \frac{\alpha_i}{L} x - \sinh \frac{\alpha_i}{L} x \right) \end{aligned} \tag{3.201}$$

and

$$\alpha_i^4 = \frac{12 \rho \omega_i^2 L^4}{E h^2} \tag{3.202}$$

In these equations (3.200–3.202), every vibration mode i is characterized by an amplitude, C_i; a radial frequency, ω_i; and a wavelength, L/α_i. Due to the vibration constraints, imposed by the boundary conditions, the parameters α_i and ω_i are *discrete*. In particular, for the singly anchored beam, for which we have,

$$\Phi(0) = 0 \quad \frac{d\Phi(0)}{dx} = 0 \tag{3.203}$$

it can be deduced that α_i is determined by solving $\cos \alpha_i \cosh \alpha_i = -1$, which leads to the following values,

$$\begin{aligned} &\alpha_1 = 1.88393 \quad \alpha_2 = 7469 \quad \alpha_3 = 7.85 \quad \alpha_4 = 11 \\ &\alpha_i = \left(i - \frac{1}{2} \right) \pi \quad \text{for } i \geq 5 \end{aligned} \tag{3.204}$$

On the other hand, for the doubly anchored beam, the boundary conditions are,

$$\begin{aligned} \Phi(0) = 0 \quad \frac{d\Phi(0)}{dx} = 0 \\ \Phi(L) = 0 \quad \frac{d^2\Phi(0)}{dx^2} = 0 \end{aligned} \tag{3.205}$$

from where it can be deduced that α_i is determined by solving $\tan \alpha_i = \tanh \alpha_i$, which leads to the following values,

$$\begin{aligned} &\alpha_1 = 3.93 \quad \alpha_2 = 7.07 \quad \alpha_3 = 10.21 \\ &\alpha_i = \left(i + \frac{1}{4} \right) \pi \quad \text{for } i \geq 4 \end{aligned} \tag{3.206}$$

Now, in analogy with the calculation for the random vibration of a particle, whose average translational energy equals $\frac{3}{2}k_BT$, in accord with the equipartition of energy, we must calculate the beam displacement energy and also apply to it the energy equipartition postulate. So, we begin by calculating the total beam energy,

$$W_T = \frac{EWh^3}{24}\int_0^L \left(\frac{d^2z}{dx^2}\right)^2 dx + \frac{\rho Wh}{2}\int_0^L \left(\frac{dz}{dt}\right)^2 dx \tag{3.207}$$

where the first term after the equal sign is its potential energy, the second term is its kinetic energy, and L is the beam length. Utilizing Equation (3.200), with $T_i \cong \sin(\omega_i t + \delta_i)$ and $T_i^* \cong \cos(\omega_i t + \delta_i)$ for the singly anchored and the doubly anchored beams, respectively, it can be shown that W_T may be expressed as,

$$W_T = \frac{EWh^3}{24L^3}\sum_{i=1}^{\infty} C_i^2\alpha_i^4 T_i^2 I_i + \frac{\rho WhL}{2}\sum_{i=1}^{\infty} C_i^2\omega_i^2 T_i^{*2} I_i \tag{3.208}$$

If we define $q_i \cong C_i T_i$ and $p_i \cong M dq_i/dt = M\omega_i C_i T_i^2$, where $M = \rho WhL$ is the total beam mass, then Equation (3.208) may be written in the form of the harmonic oscillator energy [68], that is,

$$W_T = \frac{K}{2}\sum_{i=1}^{\infty} q_i^2\alpha_i^4\frac{I_i}{3} + \frac{1}{2M}\sum_{i=1}^{\infty} p_i^2 I_i \tag{3.209}$$

which implies that the total energy is the sum of independent quadratic terms in q_i and p_i. For the cantilever beam, the total displacement, that is, including all modes, is captured by the sum of displacements at its tip, $z = \sum_i q_i\Phi_i(L)$. Since in thermal equilibrium energy equipartition requires that each term be equal to $\frac{1}{2}k_BT$, one obtains that,

$$\frac{K}{2}\overline{q_i^2}\alpha_i^4\frac{I_i}{3} = \frac{1}{2}k_BT \rightarrow \overline{q_i^2} = \frac{3k_BT}{\alpha_i^4 K I_i} \tag{3.210}$$

and, given $\overline{q_i^2}$, we can calculate the mean square displacement as,

$$\overline{z_i^2} = \overline{q_i^2\Phi_i^2(L)} = \overline{q_i^2}\Phi_i^2(L) \tag{3.211}$$

or

$$\overline{z_i^2} = \frac{k_BT}{K}\frac{3\Phi_i^2(L)}{\alpha_i^4 I_i} \tag{3.212}$$

The fraction $\Phi_i^2(L)/I_i = 4$ is obtained by [69] as follows. From Equation (3.201), given,

$$\Phi_i = (\sin\alpha_i + \sinh\alpha_i)\left(\cos\frac{\alpha_i}{L}x - \cosh\frac{\alpha_i}{L}x\right) - (\cos\alpha_i + \cosh\alpha_i)\left(\sin\frac{\alpha_i}{L}x - \sinh\frac{\alpha_i}{L}x\right) \tag{3.213}$$

substituting $x = L$ we get,

$$\begin{aligned}
\Phi_i(L) &= (\sin\alpha_i + \sinh\alpha_i)\left(\cos\frac{\alpha_i}{L}L - \cosh\frac{\alpha_i}{L}L\right) \\
&\quad - (\cos\alpha_i + \cosh\alpha_i)\left(\sin\frac{\alpha_i}{L}L - \sinh\frac{\alpha_i}{L}L\right) \\
&= (\sin\alpha_i + \sinh\alpha_i)(\cos\alpha_i - \cosh\alpha_i) \\
&\quad - (\cos\alpha_i + \cosh\alpha_i)(\sin\alpha_i - \sinh\alpha_i) \\
&= 2\cos\alpha_{\mathrm{i}}\sinh\alpha_{\mathrm{i}} - 2\sin\alpha_{\mathrm{i}}\cosh\alpha_{\mathrm{i}}
\end{aligned} \tag{3.214}$$

Then, taking the square of $\Phi_i(L)$, we get,

$$\begin{aligned}
\Phi_i^2(L) &= (2\cos\alpha_{\mathrm{i}}\sinh\alpha_{\mathrm{i}} - 2\sin\alpha_{\mathrm{i}}\cosh\alpha_{\mathrm{i}})^2 \\
&= 4(\cos\alpha_{\mathrm{i}}\sinh\alpha_{\mathrm{i}} - \sin\alpha_{\mathrm{i}}\cosh\alpha_{\mathrm{i}})^2 \\
\rightarrow \frac{\Phi_i^2(L)}{4} &= (\cos\alpha_{\mathrm{i}}\sinh\alpha_{\mathrm{i}})^2 - 2\cos\alpha_{\mathrm{i}}\sinh\alpha_{\mathrm{i}}\cdot\sin\alpha_{\mathrm{i}}\cosh\alpha_{\mathrm{i}} \\
&\quad + (\sin\alpha_{\mathrm{i}}\cosh\alpha_{\mathrm{i}})^2 \\
&= \cos^2\alpha_{\mathrm{i}}\sinh^2\alpha_{\mathrm{i}} + \sin^2\alpha_{\mathrm{i}}\cosh^2\alpha_{\mathrm{i}} \\
&\quad - 2\cos\alpha_{\mathrm{i}}\sinh\alpha_{\mathrm{i}}\cdot\sin\alpha_{\mathrm{i}}\cosh\alpha_{\mathrm{i}}
\end{aligned} \tag{3.215}$$

but, since for a singly anchored beam we have that $\cos\alpha_i\cosh\alpha_i = -1$, this equation becomes,

$$\begin{aligned}
\frac{\Phi_i^2(L)}{4} &= \cos^2\alpha_{\mathrm{i}}\sinh^2\alpha_{\mathrm{i}} + \sin^2\alpha_{\mathrm{i}}\cosh^2\alpha_{\mathrm{i}} \\
&\quad - 2\cos\alpha_{\mathrm{i}}\cosh\alpha_{\mathrm{i}}\sin\alpha_{\mathrm{i}}\sinh\alpha_{\mathrm{i}} \\
&= \cos^2\alpha_{\mathrm{i}}\sinh^2\alpha_{\mathrm{i}} + \sin^2\alpha_{\mathrm{i}}\cosh^2\alpha_{\mathrm{i}} - 2(-1)\sin\alpha_{\mathrm{i}}\sinh\alpha_{\mathrm{i}} \\
&= \cos^2\alpha_{\mathrm{i}}\sinh^2\alpha_{\mathrm{i}} + \sin^2\alpha_{\mathrm{i}}(1 + \sinh^2\alpha_{\mathrm{i}}) + 2\sin\alpha_{\mathrm{i}}\sinh\alpha_{\mathrm{i}} \\
&= \cos^2\alpha_{\mathrm{i}}\sinh^2\alpha_{\mathrm{i}} + \sin^2\alpha_{\mathrm{i}} + \sin^2\alpha_{\mathrm{i}}\sinh^2\alpha_{\mathrm{i}} + 2\sin\alpha_{\mathrm{i}}\sinh\alpha_{\mathrm{i}} \\
&= (\cos^2\alpha_{\mathrm{i}} + \sin^2\alpha_{\mathrm{i}})\sinh^2\alpha_{\mathrm{i}} + \sin^2\alpha_{\mathrm{i}} + 2\sin\alpha_{\mathrm{i}}\sinh\alpha_{\mathrm{i}}
\end{aligned}$$

$$\begin{aligned} &= (1)\sinh^2\alpha_\mathrm{i} + \sin^2\alpha_\mathrm{i} + 2\sin\alpha_\mathrm{i}\sinh\alpha_\mathrm{i} \\ &= \sinh^2\alpha_\mathrm{i} + \sin^2\alpha_\mathrm{i} + 2\sin\alpha_\mathrm{i}\sinh\alpha_\mathrm{i} \\ \frac{\Phi_i^2(L)}{4} &= (\sinh\alpha_\mathrm{i} + \sin\alpha_\mathrm{i})^2 \end{aligned} \tag{3.216}$$

But, it can be shown, making use of the orthogonality of any two modes Φ_i, Φ_j, that both for the free end beam and fixed end beam,

$$\frac{1}{L}\int_0^L \Phi_i^2 dx = (\sin\alpha_i + \sinh\alpha_i)^2 \cong I_i \tag{3.217}$$

so that, Equation (3.216) may be written as,

$$\frac{\Phi_i^2(L)}{I_i} = 4 \tag{3.218}$$

and we finally get,

$$\overline{z_i^2} = \frac{12k_BT}{K\alpha_i^4} \tag{3.219}$$

which shows that the mean square displacement of mode i is proportional to the temperature and is inversely proportional to the beam spring constant and to α_i^4. Thus, at a given temperature, the larger the spring constant and the higher the vibration mode (i.e., the larger α_i^4), the smaller the amplitude of the thermal random motion.

From the average of the square thermal displacement per mode, Equation (3.219), we proceed to obtain the total (resultant) average thermal displacement for all the modes by weighing them with Boltzmann's probability that they would be occupied,

$$P_i \propto e^{-\frac{W_i}{k_BT}} = e^{-\frac{\frac{K}{2}\sum_{i=1}^{\infty} q_i^2\alpha_i^4\frac{I_i}{3}}{k_BT}} \tag{3.220}$$

where W_i is the potential energy per vibration mode i, which is proportional to q_i^2. As a result, the total mean square displacement is,

$$\begin{aligned} \overline{z^2} &= \sum_{i=1}^{\infty}\overline{z_i^2} = \sum_{i=1}^{\infty}\frac{12k_BT}{K\alpha_i^4} \\ &= \frac{12k_BT}{K}\sum_{i=1}^{\infty}\frac{i}{\alpha_i^4} \end{aligned} \tag{3.221}$$

With the sum $\sum_{i=1}^{\infty} 1/\alpha_i^4 = 1/12$, as worked out by Rayleigh [69], we obtain that,

$$\overline{z^2} = \frac{k_B T}{K} \tag{3.222}$$

or

$$\sqrt{\overline{z^2}} = \sqrt{\frac{k_B T}{K}} \tag{3.223}$$

By a similar development, it can be shown that for the doubly anchored beam, the total mean square displacement $\overline{z^{*2}}$ is,

$$\overline{z^{*2}} = \frac{k_B T}{3K} \tag{3.224}$$

or

$$\sqrt{\overline{z^{*2}}} = \sqrt{\frac{k_B T}{3K}} \tag{3.225}$$

Examination of $\sqrt{\overline{z^2}}$ and $\sqrt{\overline{z^{*2}}}$ suggests, therefore, that the temperature, T, and the beam spring constant, K, place a *fundamental* limitation on the minimum average random/Brownian displacement attainable by a beam.

3.4 Sensing

The Internet of Things (IoT) will embody an internet-mediated interconnected network of physical sensors residing on a plethora of objects such as inertial sensors for vehicles, smart phones, gaming controllers, activity trackers, and digital picture frames [70], and the environment, to enable its ubiquitous surveillance (and, possibly, control). Therefore, the nature of the sensors that may be employed is virtually unlimited. Given the importance of low power consumption in battery-limited IoT wireless nodes, however, we will focus on addressing the fundamentals of some key low-power NEMS sensors. Those readers interested in a thorough discussion of MEMS sensors may feel free to read Senturia's book [5] and Najafi's review article [71].

3.4.1 The Accelerometer

An accelerometer is a device that detects the acceleration of a body (frame) to which it is attached. It essentially consists of a mass (the proof mass), which is suspended via a compliant structure (e.g., a beam) anchored to the frame. Its basic operation is illustrated in Figure 3.19 [71].

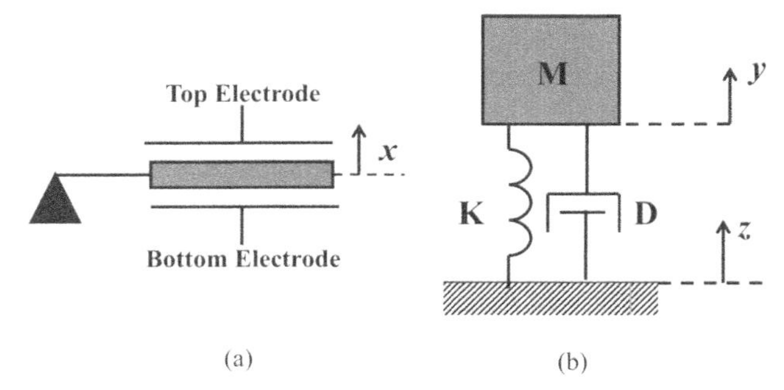

Figure 3.19 Capacitive accelerometer. (a) A proof mass is suspended between two electrodes forming a capacitor. (b) Mass-spring-dashpot model of accelerometer.
Source: Ref. [71].

When a force is applied to the frame, the proof mass moves relative to the frame. Then, the displacement of the proof mass relative to the frame is measured and related to the applied force. The system, proof mass, plus the compliant structure and damping are modeled by the second-order spring-mass-damper differential equation,

$$M\frac{d^2x}{dt^2} + D\frac{dx}{dt} + Kx = F \tag{3.226}$$

Applying the Laplace transform to this equation, one obtains the mechanical transfer function [71],

$$\frac{x(s)}{a(s)} = \frac{1}{s^2 + \frac{D}{M}s + \frac{K}{M}} = \frac{1}{s^2 + \frac{\omega_n}{Q}s + \omega_n^2} \tag{3.227}$$

where a is the external acceleration, x is the proof mass displacement, $\omega_n = \sqrt{K/M}$ is the mechanical resonance frequency, and $Q = \sqrt{KM}/D$ is the quality factor. The mechanical response of the accelerometer is a function of the frequency of the applied acceleration with respect to its resonance frequency. Thus, at low frequencies, $\omega \ll \omega_n$, Equation (3.227) becomes,

$$\left.\frac{x(s)}{a(s)}\right|_{\omega \approx 0} = \frac{M}{K} = \frac{1}{\omega_n^2} \Rightarrow x = \frac{a}{\omega_n^2} \tag{3.228}$$

This implies that, at low frequencies, the displacement depends on the ratio of the proof mass to the spring constant, not on their separate values. On the other hand, high-resonance frequencies will give rise to fast, but small, displacements, while low-resonance frequencies will give rise to large, but slow, displacements [5, 71]. Now, as discussed in the context of the laser-driven cantilever beam, any mechanical system is exposed to Brownian noise, which contributes an additional force component with mean-square spectral density given by $\sqrt{4k_BTD}$. This will result in an additional mean-square acceleration given by [5],

$$a_{noise\_rms} = \sqrt{\frac{4k_BT\omega_n}{MQ}} \tag{3.229}$$

Thus, due to Brownian noise, there will always exist an error in determining the acceleration of an object.

3.4.1.1 Capacitive accelerometer implementation

In the capacitive accelerometer, the proof mass plays also the role of one of the plates of a capacitor. Thus, as it moves closer to, or away from, the other plate, in response to the applied force, an inter-plate gap or plate-to-plate area of overlap varies, and so does the related capacitance (Figure 3.20). As shown previously, the effected capacitance change is inversely proportional to the square of the separation between the capacitor plates.

From the fact that the capacitor charge is given by,

$$Q = C(x, y, z)V \tag{3.230}$$

and that its current, under an applied voltage $V(t)$, is,

$$i(t) = \frac{dQ}{dt} = V\left(\frac{\partial C}{\partial x}\frac{dx}{dt} + \frac{\partial C}{\partial y}\frac{dy}{dt} + \frac{\partial C}{\partial z}\frac{dz}{dt}\right) + C(x, y, z)\frac{dV}{dt} \tag{3.231}$$

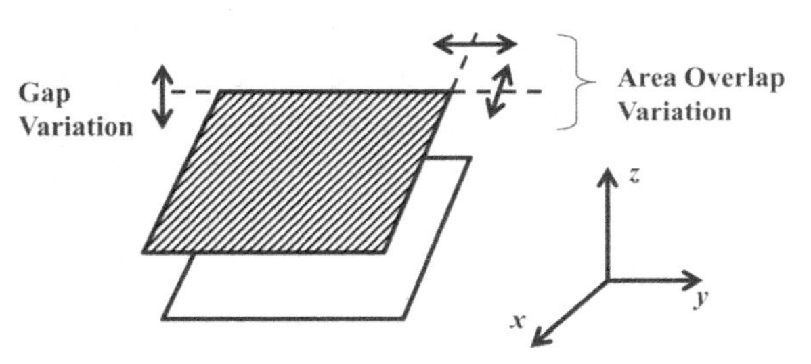

Figure 3.20 Capacitive accelerometer implementation sketch: The motion of the proof mass coincides with that of the plate of a capacitor, so that by measuring the change in area overlap or parallel-plate gap and the concomitant capacitance change, one can determine the proof mass' acceleration.

Source: Ref. [5].

it is clear that one can relate the change in capacitance due to plate displacement in x, y, and z (resulting from an acceleration) and the displacement time rate of change. In the simplest case, one would have,

$$i(t) = V\left(\frac{\partial C}{\partial x}\frac{dx}{dt}\right) + C(x)\frac{dV}{dt} \tag{3.232}$$

for an x-directed acceleration. Many circuit techniques are utilized in practice to transduce the measured acceleration into an electronic signal [5]. The transimpedance amplifier shown in Figure 3.21 exemplifies one of these.

3.4.1.2 Quantum mechanical tunneling accelerometer

The term "quantum mechanical tunneling" refers to the propagation of a quantum particle through a potential barrier where the particle's energy is lower than the barrier height (Figure 3.22) [68, 72]. In a tunneling accelerometer, the magnitude of a tunneling current that varies exponentially with the separation between a tunneling tip and its counter electrode is utilized to measure acceleration. A tunneling accelerometer typically offers better sensitivity since relatively small acceleration variations produce relatively larger responses in the exponentially responding tunneling accelerometers as compared to square-power-responding capacitive accelerometers.

A common architecture for tunneling accelerometers involves placing the tunneling tip on a cantilever end portion (proof mass); these elements

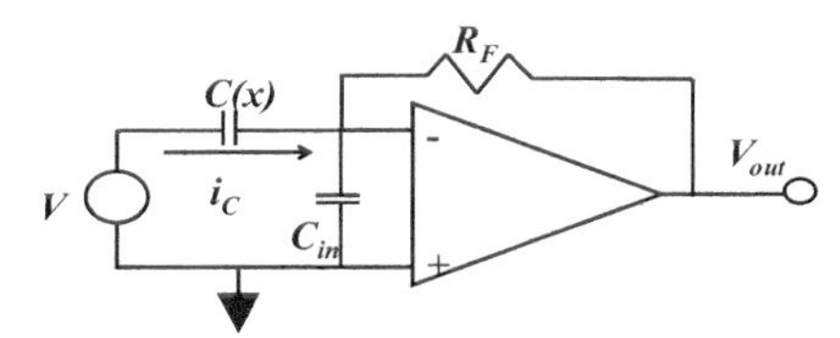

Figure 3.21 Transimpedance amplifier typifies electronic circuit to measure capacitance current from accelerometer.

Source: Ref. [5].

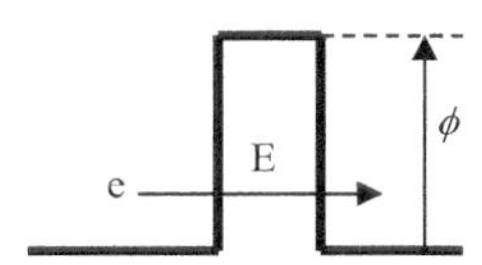

Figure 3.22 Sketch of electron e of energy E tunneling through potential barrier of height Φ.

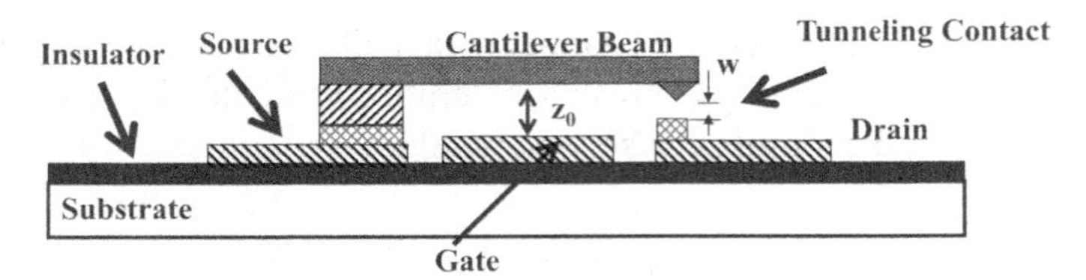

Figure 3.23 Cross-section of micromechanical tunneling transistor showing basic device structure.

Source: Ref. [73].

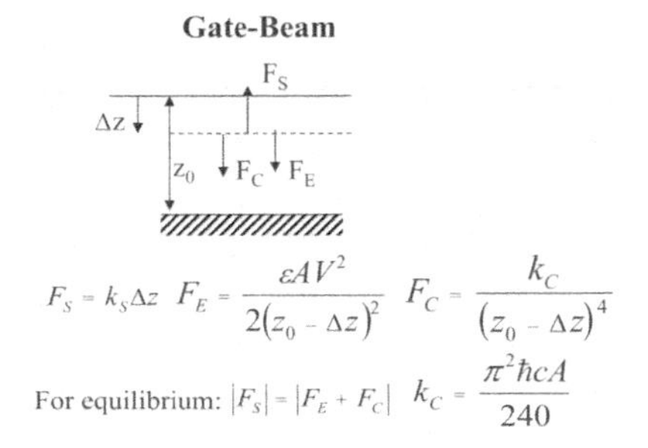

Figure 3.24 Micromechanical transistor physics: Gate-beam actuation; $\hbar$ is Planck's constant divided by 2π, and c is the speed of light.

Source: Ref. [74].

are captured in Figure 3.23 in a device called "micromechanical tunneling transistor" [73].

The tunneling current in Figure 3.23 is elicited by applying a pre-pull-in gate-beam actuation voltage that causes the fine tunneling tip to displace a distance Δz, which *reduces* the physical width w of the barrier. In particular, the tunneling current produced as the cantilever beam deflects is given by,

$$I_{Tun} = V_{ds} K \exp(-1.025\sqrt{\Phi}(w - \Delta z)) \tag{3.233}$$

where Φ and w are the tunneling barrier height and width, respectively [73]. The nominal barrier width (w) is reduced by the deflection Δz of the beam (Figures 3.24 and 3.25), which results from the application of the voltage $\mathrm{V} = \mathrm{V_g} + \mathrm{V_{ds}}$, the sum of the gate-beam and source-drain voltages. Δz is obtained, in turn, from an equation for the balance of the forces determining the cantilever beam equilibrium position, namely the electrostatic F_E, spring

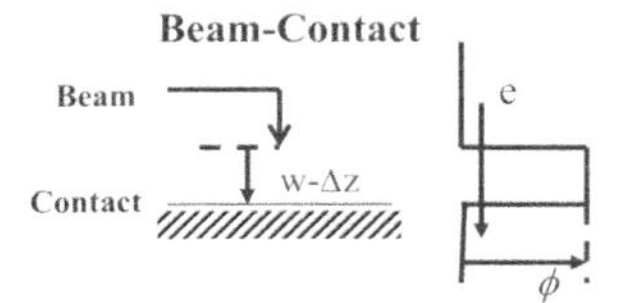

Figure 3.25 Micromechanical transistor physics: Beam-contact tunneling.

Source: Ref. [74].

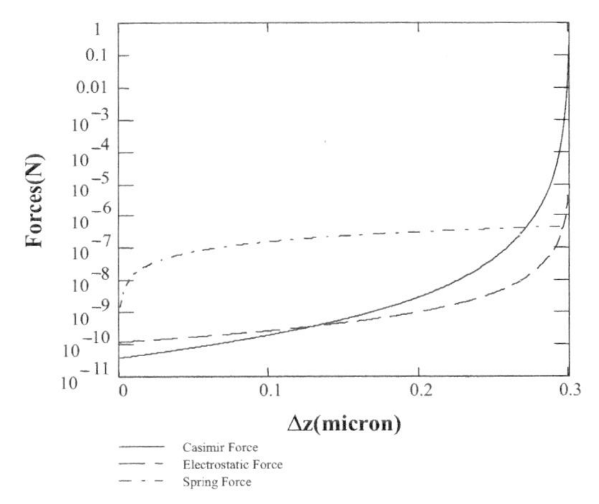

Figure 3.26 Typical relative variations and magnitudes of Casimir, electrostatic, and spring forces in micromechanical tunneling transistor.

Source: Ref. [74].

F_S, and Casimir forces F_C (Figure 3.26) [74],

$$\frac{\pi^2 \hbar c A}{240(z_0 - \Delta z)^4} + \frac{\varepsilon A V^2}{2(z_0 - \Delta z)^2} - k_S \Delta z = 0 \tag{3.234}$$

In practice, in response to the applied acceleration, a proof mass responds by moving closer or further away from the tunneling tip. The flexibility of the cantilever is exploited through the application of a bias voltage between the cantilever end and one or more biasing electrodes to flex the cantilever appropriately so that the tunneling tip is within the tunneling range of the counter electrode. Note, however, that since the proof mass moves orthogonally to the cantilever longitudinal axis (i.e., either towards or away from the cantilever), there is always the danger of a sufficiently strong acceleration

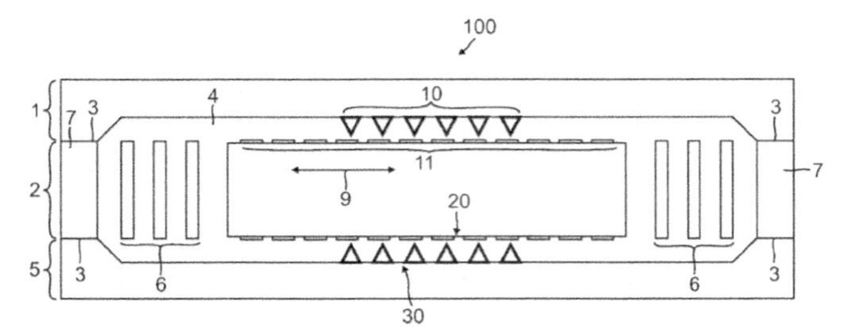

Figure 3.27 MEMS tunneling accelerometer [75, 76].

causing the tunneling tip to contact the counter electrode on the proof mass. Since the tunneling tip dimensions at the tip apex are typically on the order of just a few atoms, such a contact could readily damage the tunneling tip. Thus, stops, or other means, are required to prevent the contact, which decreases the achievable measurement range. In addition, the sensitivity of conventional tunneling architectures is limited by the *single* tunneling tip.

A robust, tunneling accelerometer that overcomes the above sensitivity limitations is shown in Figure 3.27 [75, 76].

In this accelerometer concept, the tunneling tips 10 are disposed as an array perpendicular to the interior of capping wafers 1 and 5, and the counter electrodes 11 are patterned on the proof mass 9 attached to suspensions 6, the cap wafers and proof mass being bonded at surfaces 3. In this scheme, the proof mass moves orthogonally to the tunneling tips so a constant distance between them and the counter electrodes is maintained, thus avoiding the possibility of the tips crashing on the counter electrodes; in the absence of lateral acceleration, an applied bias voltage causes the tunneling current to flow between tips and counter electrodes due to their alignment. As the proof mass moves, in response to a lateral acceleration, the counter electrodes misalign with respect to the tunneling tips and the total current changes. This change is proportional to the acceleration and thus serves to measure it.

3.4.2 Vibration Sensors

While we encounter *inertial* sensors frequently in our daily lives, *vibration* sensors, aimed at monitoring the health of expensive industrial machinery, embody a crucial function in major factories [77, 78]. Vibration sensors provide an easy, cost-effective means of monitoring and protecting critical machinery on a continuous basis by enabling [77]. Typical applications include critical pumps and motors, cooling towers and fans, slow speed rolls, and rotary and screw compressors.

The key insight that allows monitoring the health of a machine by sensing its vibrations (oscillatory motion) is that fatigue of the moving members of a machine is related to the number of cycles a machine experiences before failure [78]. In particular, as characterized by the Wöhler curve, describing the endurance strength, that is, the stress level up to which a structure can be loaded a certain number of times, at high stresses, the load can only be carried a few times; however, reduction of the stress increases the number of cycles to failure. On the other hand, it has been determined that it is possible to find a stress level such that, if operated below this level, the endurance of a structure becomes infinite.

In general, the vibrations in a machine are the result of [77] (i) an imbalance in the forces applied; (ii) shock forces; (iii) frictional forces; and (iv) acoustic forces. These forces, as previously shown, are related to the machine's structural parameters, namely its mass, its stiffness, and its damping.

Measuring vibration throughout the machine enables the following steps to safeguard its health [77]:

(i) Verification of whether or not the frequency amplitudes produced exceed the material limits
(ii) Avoiding the excitation of resonances at certain locations of a machine
(iii) Identifying where to introduce damping or isolate the vibration sources of resonance
(iv) Undertaking in a timely fashion preventive maintenance measures on machines
(v) Developing computer models of a machine component.

Clearly, once the acceleration is determined with an accelerometer, the velocity and displacements may be determined, given that if the acceleration is given by,

$$a = A \sin \omega t \tag{3.235}$$

then the velocity is given by,

$$v = \int a dt = -\frac{A}{\omega} \cos \omega t \tag{3.236}$$

and the displacement is given by,

$$d = \iint a dt dt = -\frac{A}{\omega^2} \sin \omega t \tag{3.237}$$

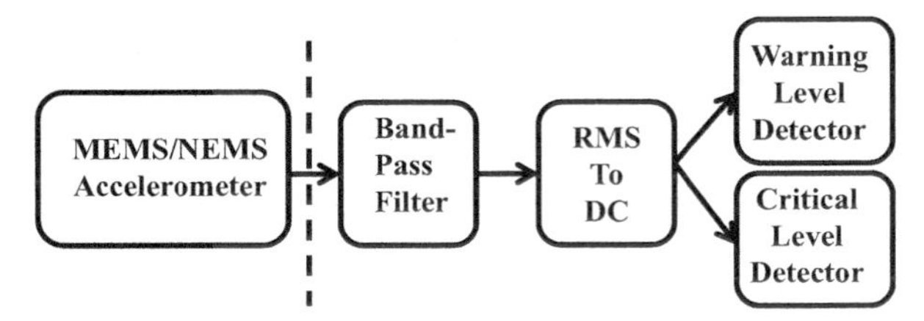

Figure 3.28 Time-domain vibration signal chain example.

Source: Ref. [77].

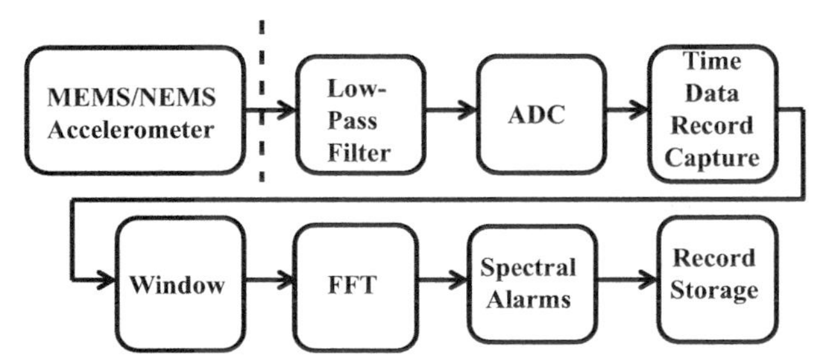

Figure 3.29 Example of signal chain for spectral vibration analysis. ADC: Analog-to-digital converter, FFT: fast Fourier transform.

Source: Ref. [77].

where $\omega = 2\pi f$ is the radial frequency and f is the frequency in Hz. Machine dynamics may be derived as per the electronic systems architectures shown in Figures 3.28 and 3.29.

A vibration sensing system may be partitioned as shown in Figures 3.28 and 3.29. In other words, the accelerometer may be packaged alone and endowed with an interface to the outside world, or packaged together with the electronics/intelligence. Other reasons such as environmental parameters, available size, and available power may be decided in determining sensor implementation.

In summary, vibration sensors are essential to effect preventive maintenance and avoid the occurrence of a costly catastrophic failure. Furthermore, it is clear that connecting *wireless* vibration sensors to machines, and linking them via IoT to computers, allows algorithms (machine learning, artificial intelligence) to predict problems based on the sound a machine makes [77–79].

3.5 Summary

In this chapter, we have dealt with the fundamental physics on which nano-electromechanical quantum circuits and systems are based. We began by presenting the actuation mechanisms which we estimate of greatest importance and relevance for MEMS/NEMS devices in the context of the IoT, namely electrostatic, piezoelectric, Casimir (quantum electrodynamical), and radiation pressure. This was followed by a discussion of mechanical vibration and its computation for fundamental structures, in particular, the cantilever and doubly anchored beams. Then, since the temperature impacts the purity of the mechanical resonance frequency of mechanical structures, due to the temperature-induced Brownian motion, we set up to re-derive historically fundamental equations relating the random vibration to the dissipation germane to a mechanical structure. We concluded the chapter by presenting fundamental sensing approaches, namely those for acceleration, velocity, and position, exploiting capacitance or quantum mechanical tunneling current variation.

4

Understanding MEMS/NEMS Devices

4.1 Introduction

In this chapter, we deal with microelectromechanical system (MEMS)/nano-electromechanical system (NEMS) devices [5, 6] that have high potential to be employed on Internet of Things (IoT) applications, namely switches, varactors, and resonators [10]. These devices are key building blocks for low-power consumption front-end wireless transceivers [81, 82] of the type that may be employed in IoT nodes. In particular, we discuss their structure and principles of operation and provide a summary of recent performance results.

4.2 MEMS/NEMS Switches

Switches are electrical devices utilized to effect signal routing and signal blocking or to turn power ON/OFF [10]. The fundamental MEMS/NEMS switch topologies and their biasing schemes are shown in Figures 4.1–4.6.

In one implementation of a radio-frequency (RF) MEMS switch, it consists of a metallic micromechanical beam disposed over and transversal to a coplanar waveguide (CPW) transmission line so that, when the beam is not being actuated, that is, when it is in an un-deflected configuration, a signal propagating down the CPW line passes the location of the beam with minimal attenuation, but when actuated the beam contacts a thin dielectric layer covering the signal line. In this case, a relatively large capacitance is formed by the beam–dielectric–signal line "sandwich", RF-wise at this location, such that the propagating signal perceives a low impedance to the ground plane strips of the CPW, thus substantially blocking transmission by reflecting an incoming propagating signal. Figure 4.1 shows the end views of this switch type in both the passing and blocking states as well as a top view of it. In this implementation, the signal carries both the RF signal and the beam actuation DC control bias. Because the beam shunts the CPW line

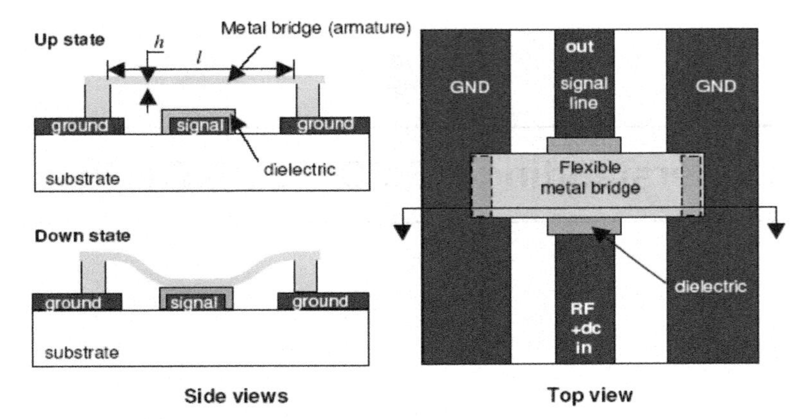

Figure 4.1 Configuration and biasing of RF MEMS switch structure: Electrostatically actuated capacitive shunt switch (broadside configuration) implemented on a CPW transmission line. © [2004] IEEE. Reprinted with permission from De Los Santos et al. [82].

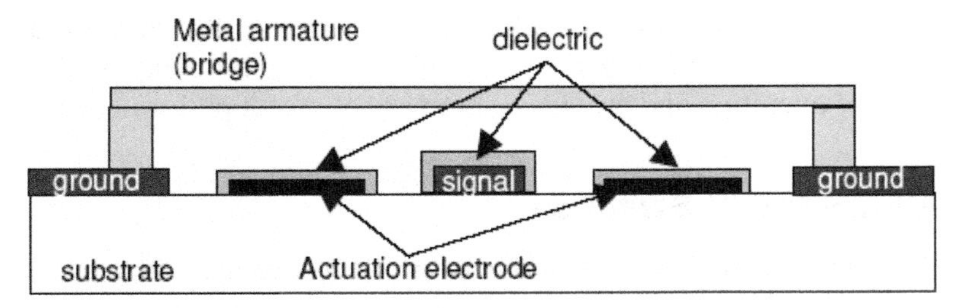

Figure 4.2 Configuration and biasing of RF MEMS switch structure: Electrostatically actuated three-terminal capacitive shunt switch (in a broadside configuration). © [2004] IEEE. Reprinted with permission from De Los Santos et al. [82].

when actuated, this type of switch is often referred to as a "shunt" switch. In a second variation of this switch, the DC voltage is applied at electrodes that are separate and distinct from the signal line; this is shown in Figure 4.2. These switches are called "capacitive" because the contact between the metallic beam and the CPW signal line is mediated by a dielectric, and thus, the location of the contact is a capacitor.

In another implementation of an RF MEMS switch, the beam is disposed so that when not actuated it extends over and above a gap in a transmission line, where this gap reflects the signal propagating down the line. When the beam is actuated, however, it bridges said gap, thus enabling signal propagation with minimal attenuation. In this implementation, the beam may

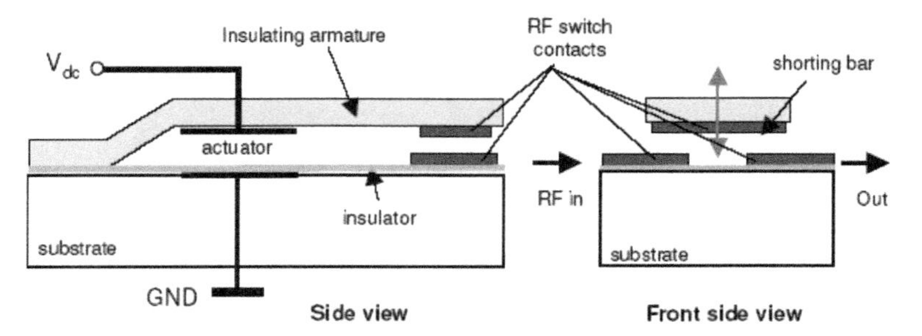

Figure 4.3 Configuration and biasing of RF MEMS switch structure: Electrostatically actuated ohmic series relay implemented with a cantilever beam armature (broadside configuration). © [2004] IEEE. Reprinted with permission from De Los Santos et al. [82].

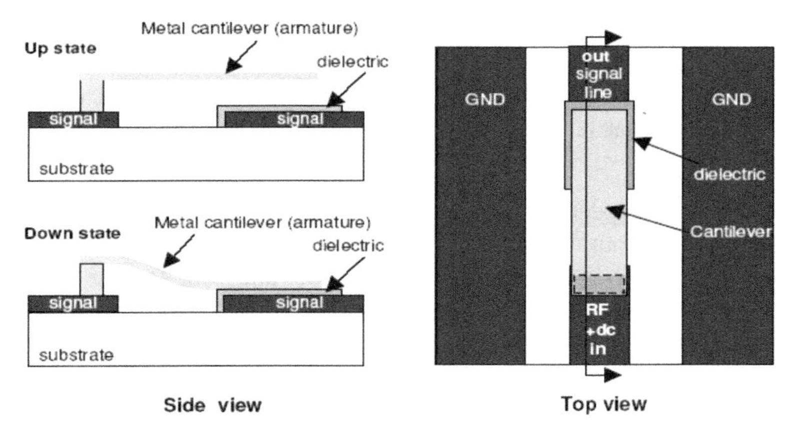

Figure 4.4 Configuration and biasing of RF MEMS switch structure: Electrostatically actuated capacitive series switch implemented on a CPW transmission line (in-line configuration). © [2004] IEEE. Reprinted with permission from De Los Santos et al. [82].

be oriented either perpendicular to the transmission line (Figure 4.3) or in-line (in "series") with it (Figure 4.4).

In the configuration of Figure 4.3, it is noticed that the beam is made up of a non-metallic material, but under its tip it has a metal contact attached so that there is no dielectric between the beam and the transmission line at the points of contact. This type of switch is often referred to as an "ohmic" switch.

RF MEMS switches are variously called as shunt capacitive or ohmic switches or series capacitive or ohmic switches [83].

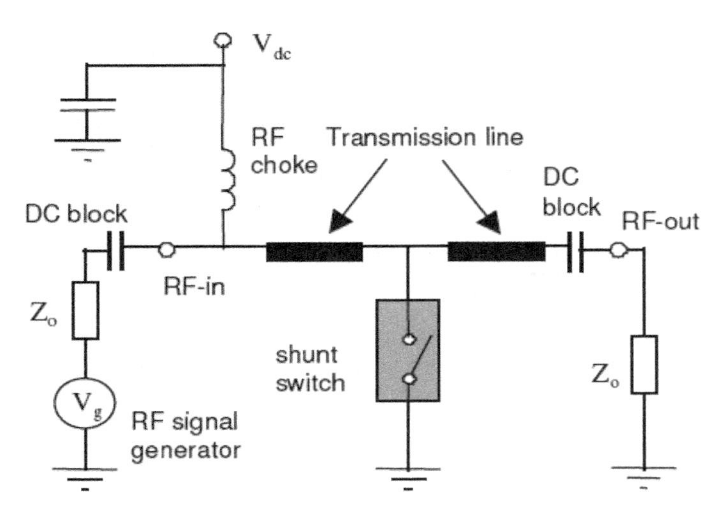

Figure 4.5 Biasing network of shunt switch. © [2004] IEEE. Reprinted with permission from De Los Santos et al. [82].

An important aspect or question in operating RF MEMS switches is "How to bias them without interfering with their RF/microwave performance?" This is answered conceptually in Figures 4.5 and 4.6, for the shunt and series switches, respectively. The general idea that the diagrams intend to convey is that the DC voltage must be isolated from the high-frequency signal path by a high impedance. This high impedance is represented by the "RF choke." In addition, the signal source and load must be isolated from the DC via "blocking capacitors." In practice, where the bias voltage is dynamic/pulsed, the "driver circuit," used to activate/deactivate the switches, must be carefully designed to drive the switching at the desired frequency, yet not disturb their RF/microwave performance. For a comprehensive work on RF MEMS, including many practical application issues, the reader is referred to Lucyszyn's edited book [83] and a recent review article [84].

4.2.1 Nanoelectromechanical Switches

The key attributes of RF MEMS switches reside in their ability to operate at the highest signal frequencies, while exhibiting superior insertion loss, isolation, and bandwidth compared to traditional RF/microwave switches, for example, pin diode-based or field-effect transistor (FET)-based switches [83]. Research and development pertaining to RF MEMS switches has revolved

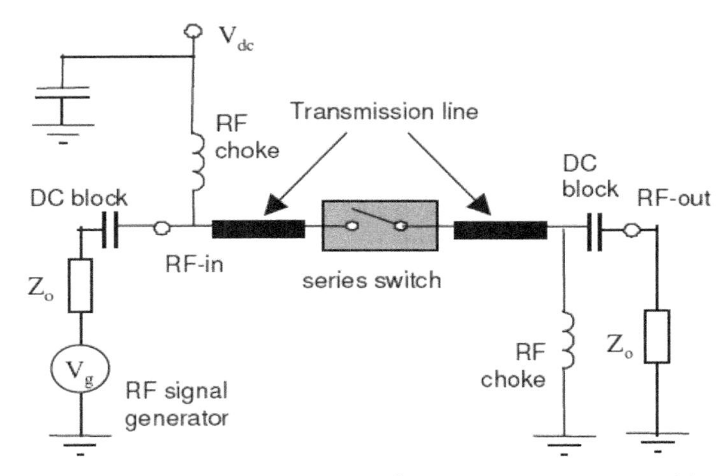

Figure 4.6 Biasing network of series switch. © [2004] IEEE. Reprinted with permission from De Los Santos et al. [82].

around effecting implementations in a variety of frequency bands and several reliability issues, such as dielectric charging, contact failure, and temperature instability [83]. In particular, methods to improve the performance of MEMS switches at high signal frequency include the following [84]: (i) combining switches in various circuit topologies such as several switches in series, in a "T" – or a π-network; (ii) using dielectric layer materials with high dielectric constants and conductor materials with low resistance; (iii) employing various fabrication process implementations such as MEMS switches based on bipolar complementary metal–oxide–semiconductor (BiCMOS) technology; (iv) employing thermal compensation structures, circularly symmetric structures, thermal buckle-beam actuators, molybdenum membrane, and thin-film packaging; (v) selecting synthetic diamond or aluminum nitride dielectric materials and applying a bipolar driving voltage, stoppers, and a double-dielectric-layer structures; and (vi) adopting gold alloying with carbon nanotubes (CNTs), hermetic and packaging, and mN-level contacts.

With the emergence of nanotechnology, one question that comes to mind is "How could this novel technological approach to miniaturize things be exploited to improve the performance of RF MEMS switches?" [72]. The answer lies in at least two realms, namely downscaling and novel materials, which are addressed subsequently.

4.2.1.1 Downscaled MEMS/NEMS switches

As is well known, the switching speed, that is, the rapidity with which traditional RF MEMS switches can be turned ON/OFF, is limited to several microseconds [83]. This limitation is rooted in the large dimensions and, concomitantly, mass, which these switches possess. For instance, a typical RF MEMS switch design has a beam with area in the neighborhood of 200 μm × 200 μm, a thickness of 2 μm and a beam-to-bottom electrode distance of about 2.5 μm. Reducing the switch mass is an obvious course of action to increase its switching frequency, given that the switch resonance frequency is inversely proportional to the square root of its mass, that is

$$f_0 = \frac{1}{2\pi}\sqrt{\frac{k}{m}} \tag{4.1}$$

Similarly, the strengthening of the MEMS beam to increase its stiffness (spring constant, k) should be pursued. In fact, an expression for the switching time of a MEMS switch that takes into consideration its pull-in voltage, V_p; the applied voltage, V_{app}; and its resonance frequency, namely

$$t_s = 3.67\frac{V_p}{2\pi V_{app} f_0} \tag{4.2}$$

suggests that, for a given V_p, increasing V_{app} and f_0 leads to a reduction in switching time [85].

The approach undertaken by Verger et al. [85], who demonstrated a sub-hundred nanoseconds miniaturized RF MEMS switch [85], was to increase the overall beam mechanical stiffness, simultaneously keeping its effective mass as low as possible. In particular, they determined via multiphysics 3D finite-element simulations that a beam design integrating stacked layers of gold, aluminum, and alumina with their respective Young's modulus of 78 GPa, 70 GPa, and 380 GPa would allow them to attain the desired stiffness and, consequently, switching speed. For a variety of beam sizes and stacked layer thicknesses, the resonance frequency was obtained and plugged into Equation (4.2) until the desired switching speed was attained. The final dimensions were a length of 25 μm and a width of 15 μm, and the beam layer thicknesses of alumina/aluminum/alumina stack were 100 nm/250 nm/100 nm for a switching speed of 51 ns. A distance of 300 nm between the beam and bottom electrode was chosen to obtain an actuation voltage between 20 V and 60 V.

The fabrication of the miniaturized beam followed the technique of *surface micromachining* [10], in which the wafer has thin film materials selectively added to and removed from it. The film materials that are eventually removed are called *sacrificial materials*, whereas those that remain are called *structural materials*. Figure 4.7 shows the process flow for making the switches.

First (Figure 4.7a), a metallization layer consisting of titanium (10 nm)/gold (100 nm)/titanium (10 nm) is evaporated and patterned on a fused silica substrate to define the beam actuation electrode and the coplanar-waveguide (CPW) RF transmission line. This substrate was chosen because of its low loss, necessary for good (low insertion loss) RF performance. Then (Figure 4.7b), a dielectric layer for the isolation/protection of the bottom electrode, consisting of a 400-nm-thick layer of alumina, was deposited by plasma-enhanced chemical vapor deposition (PECVD). This was followed (Figure 4.7c) by the deposition and patterning of a 300-nm-thick photoresist sacrificial layer (this layer was to be removed at the end of the process) to define the suspended structures, for example, beam and areas. Next (Figure 4.7d, e), the multilayer beam was deposited using the "lift-off" technique. In this process, openings are made in the photoresist by UV photolithography to define beam geometries. Then, the composite structural material is deposited using a pulsed UV laser that causes it to evaporate upon being heated by the impinging laser light onto sapphire and pure aluminum targets. The beam fabrication concludes by lifting them off (Figure 4.7f) by submersion in acetone and depositing a titanium (10 nm)/gold (150 nm) layer followed by its subsequent electroplating up to 1 μm thick in order to anchor the beam and thicken the RF line (Figure 4.7g). The last step (Figure 4.7h) is to release the beams by wet etching followed by drying in a CO_2 critical point dryer.

How was the fabricated beam performance tested? Two aspects of the device were tested, namely its mechanical resonance frequency and its switching time.

The fundamental mechanical resonance frequency was determined using the set up shown in Figure 4.8.

Here, an RF signal drives the input of the beam, while it is actuated by a sinusoidal signal which, in turn, modulates the amplitude of the input signal as it executes a periodic motion that modulates its capacitance at the actuation signal frequency. The transmitted signal at the beam output is then examined by a spectrum analyzer, where its frequency spectrum is displayed, and an oscilloscope, where its time waveform is displayed. Figure 4.9 shows a sketch of the spectrum analyzer display. The spectrum shows the input (carrier) signal and four additional "sideband" signals. Of these four extra frequencies,

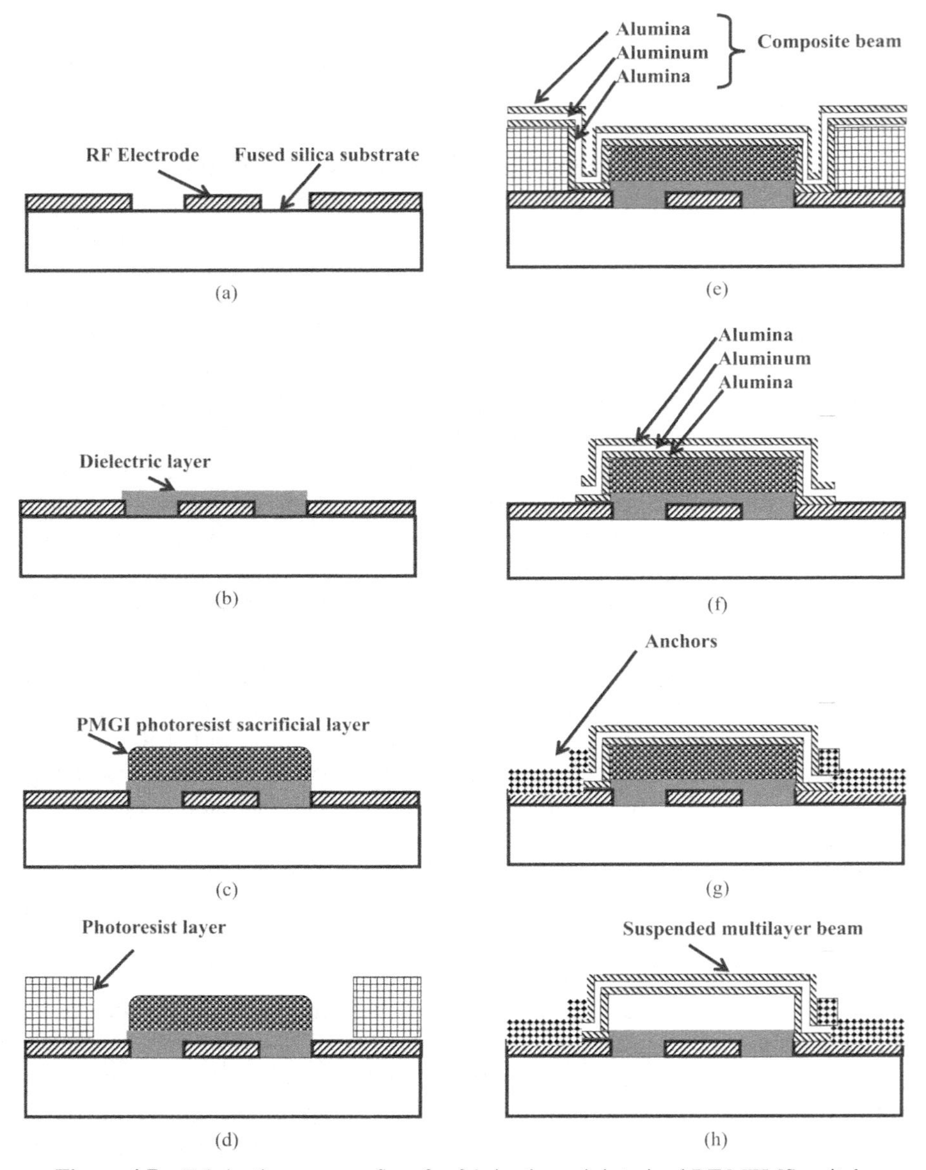

Figure 4.7 Fabrication process flow for fabricating miniaturized RF MEMS switch.
Source: Ref. [85].

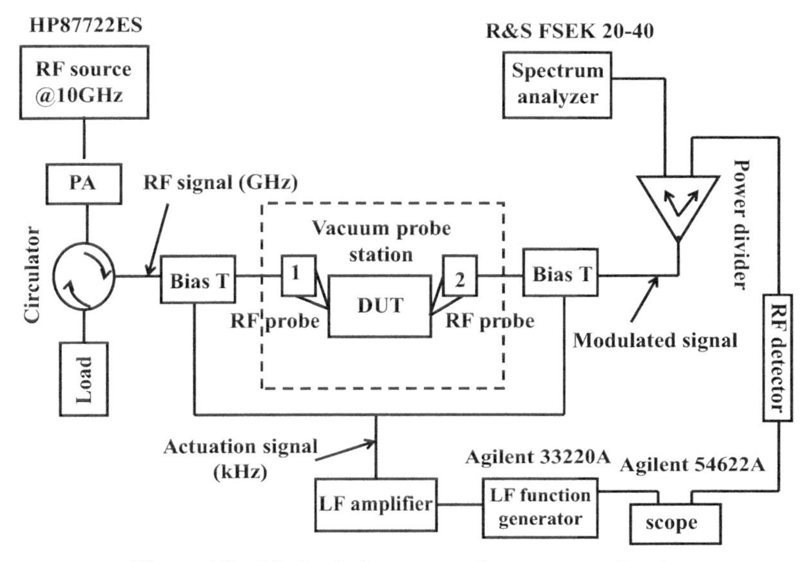

Figure 4.8 Mechanical resonance frequency test bench.

Source: Ref. [85].

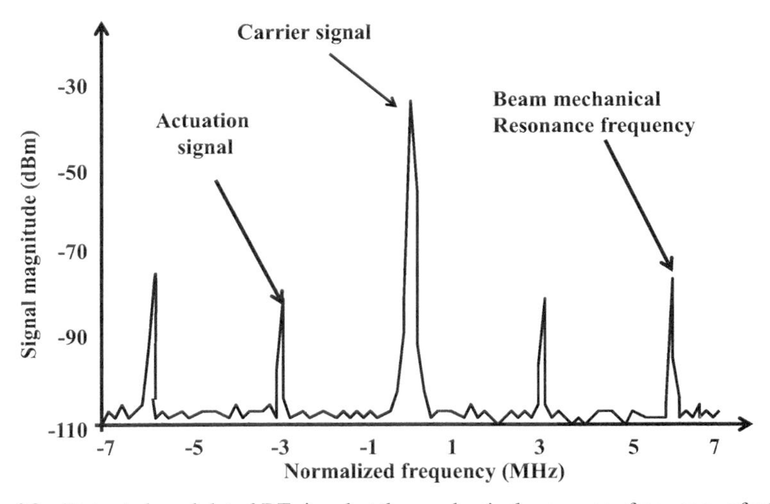

Figure 4.9 Detected modulated RF signal at the mechanical resonance frequency of a 25 μm long and 15 μm wide bridge.

Source: Ref. [85].

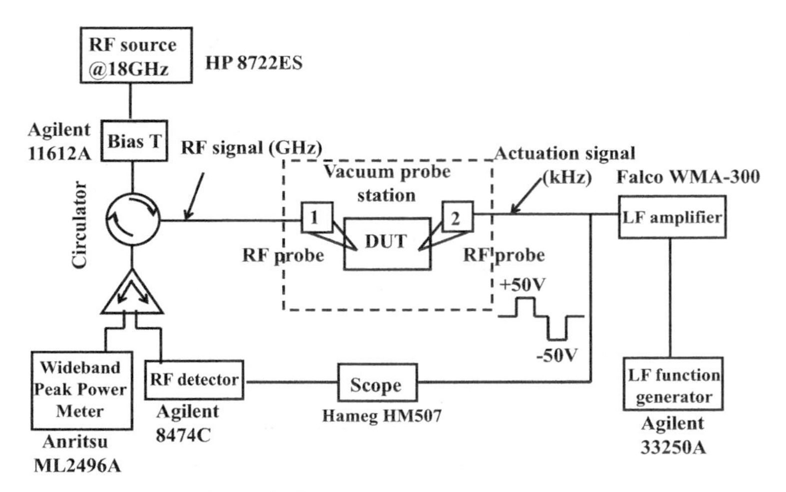

Figure 4.10 Switching speed test bench.

Source: Ref. [85].

the two closest to the carrier correspond to the actuation signal and the other two (farthest) signals correspond to the beam motion frequency. For beam motion at its mechanical resonance frequency, the amplitude of these signals is highest, as a result of which the capacitance change and signal modulation are highest. The measured mechanical resonance frequencies ranged from 6.48 MHz to 6.90 MHz, which was in good agreement with the 6.825 MHz predicted by ANSYS. The device/beam under test (DUT) was placed in a vacuum chamber to establish low damping conditions. The switching time was measured with the set up shown in Figure 4.10. Here, an RF signal drives the input of the beam, while it is actuated by a pulsed bipolar signal (the bipolar signal limits the propensity for dielectric charging).

The beam deflection amplitude versus time, resulting from the applied pulsed actuation waveform, is detected by measuring the amplitude of the input signal that is reflected. Upon receiving the reflected signal by a circulator, which directs it to a diode detector, the beam displacement is displayed in an oscilloscope, where the beam switching waveform is shown together with the actuation signal. Measurements of the switching speed were conducted for various actuation voltages (Figure 4.11).

As seen, the switching speed increases as the actuation voltage increases beyond the pull-in voltage of 30 V. In particular, the measured switching delay decreases from the order of microseconds for an actuation voltage of 30 V to 50 ns above 70 V.

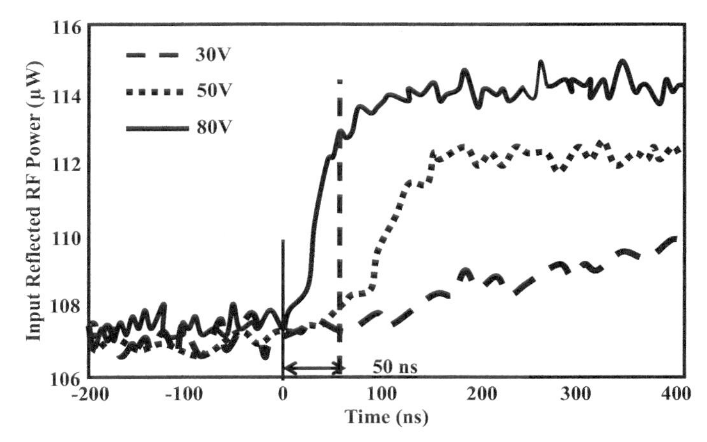

Figure 4.11 Measured switching speed of a 25 μm long and 15 μm wide beam.
Source: Ref. [85].

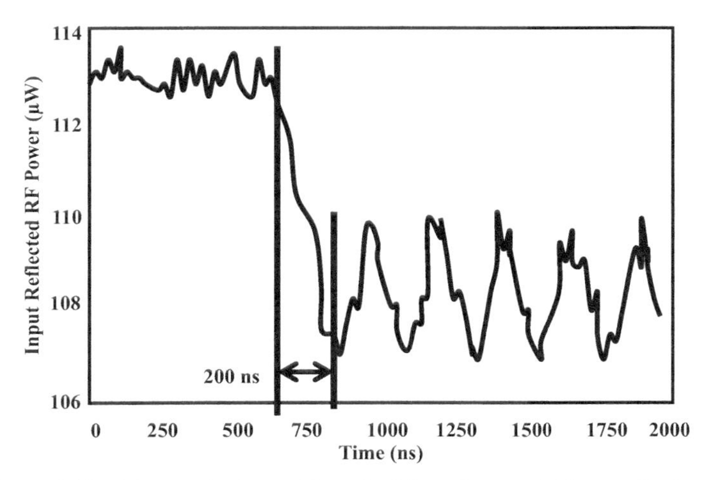

Figure 4.12 Measured release delay of a 25 μm long and 15 μm wide beam.
Source: Ref. [85].

Once the actuation voltage becomes zero, the switch is released to go back to its equilibrium position. The release time delay, shown in Figure 4.12, is of 200 ns.

It is pointed out by Verger et al. [85] that the delay time is a function of the beam/dielectric adhesion forces and that the oscillations observed upon beam release are the result of lack of air damping in the vacuum environment of the chamber housing the beam.

4.2.1.2 MEMS/NEMS switches via new materials

The properties of graphene for NEMS applications have attracted much interest [86–90]. Studies have shown that it can sustain current densities five orders of magnitude larger than ordinary metals and exhibit very large thermal conductivity and strength, in particular, a room-temperature thermal conductivity of $\sim$5000 $\mathrm{Wm^{-1}\,K^{-1}}$, a breaking strength of 42 N/m, and values of Young's modulus close to 1 TPa [86]. A remarkable property of graphene is that its mechanical behavior is also strongly influenced by the van der Waals forces, to the point that these may be exploited to clamp few layers of graphene sheets to the substrates [89, 90].

An example of a straightforward application of graphene to CPW NEMS switches was presented by Sharma et al. [91] (Figure 4.13). In this switch, the substrate consists of a high-resistivity material such as alumina or high-resistivity silicon, the signal and ground stripes consist of a high conductivity metal such as gold, and the beam consists of a graphene membrane.

The role of the graphene is that of implementing a beam that will exhibit a low actuation voltage and a fast switching speed and be amenable to monolithic integration with future all-graphene transceivers [91]. As usual, the role of the dielectric is to isolate the beam from the signal line. The graphene beam, however, behaves differently from the usual metallic beams, in that the application of a bias/actuation voltage between it and the signal line causes its conductivity to increase as a consequence of the "doping" effect it experiences when influenced by an electric field [92].

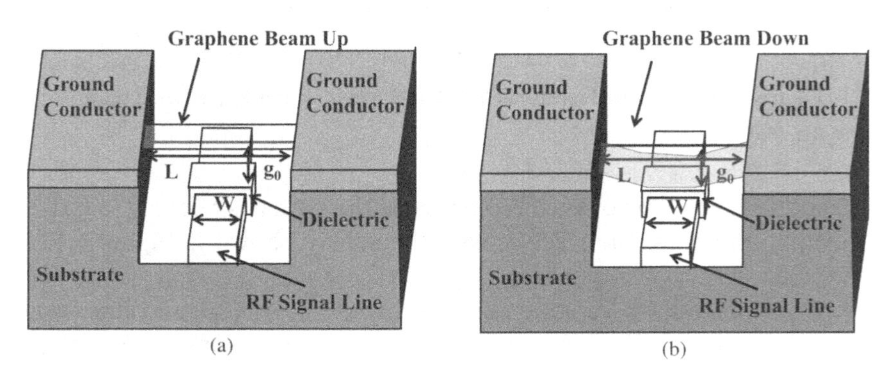

Figure 4.13 Sketch of a graphene-based RF NEMS switch in (a) up-state and (b) down-state. *Source*: Ref. [91].

In designing (and modeling) the graphene-based switch, it is necessary to capture the dependence of its conductivity with bias in both its undeflected (unbiased) and deflected (biased) configurations. This is given by the following equation, derived by Hanson et al. [93],

$$\sigma(\omega,\mu,\Gamma) = -j\frac{q_c^2 k_B T}{\pi\hbar^2(\omega - j\Gamma)}\left(\frac{|\mu_c|}{k_B T} + 2\ln(e^{-\frac{|\mu_c|}{k_B T}} + 1)\right) \tag{4.3}$$

where ω is the angular frequency, Γ is the scattering rate (its inverse $\tau = 1/\Gamma$ is the relaxation time), $T = 300$ K is the absolute temperature, $\hbar$ is the Planck constant divided by 2π, k_B is Boltzmann's constant and μ_c is the chemical potential.

The chemical potential for the unbiased graphene beam is given by the following equations [93],

$$\mu_{c\_unbiased} = \begin{cases} sign(n_{e\_unbiased} - n_{h\_unbiased})\hbar\nu_f \\ \qquad \times\sqrt{|n_{e\_unbiased} - n_{h\_unbiased}|\pi} & (4.4a) \\ \dfrac{\pi\hbar^2}{2m^*}(n_{e\_unbiased} - n_{h\_unbiased}) & (4.4b) \end{cases}$$

where $\nu_f = 10^6$ m/s is the Fermi velocity in graphene and m^* is the effective mass in multilayer graphene; $m^* = 0.052\ m_e$ (for 3,4 layers) [94], and m_e is the electron effective mass.

When the bias voltage causes a pull-in of the graphene beam, that is, $|V_{bias}| \geq |V_{pull-in}|$, the part of the graphene beam directly above the signal line experiences the electric field from it, and as a result, its carrier density in this region can be estimated from the charge balance relationship [95], namely,

$$(V_{bias} - V_{Dirac})C_{dielectric} = q(n_{e\_deflected} - n_{h\_deflected}) \tag{4.5}$$

where q is the elementary charge, and $n_{e\_deflected}$ and $n_{h\_deflected}$ are the electron and hole carrier densities in the deflected configuration. The parameter position, $C_{dielectric}$, is the capacitance of the dielectric isolating the graphene from the signal line. For an oxide dielectric, we have, $C_{dielectric} = C_{ox} = \varepsilon_r\varepsilon_0/t_{ux}$, where ε_r is the relative permittivity of the dielectric, ε_0 is the permittivity of vacuum t_{ox} is the thickness of the oxide dielectric, and V_{Dirac} is the bias voltage at the Dirac point. The Dirac point is defined by Tan et al. [95], $V_{Dirac} = -q(n_{e\_deflected} - n_{h\_deflected})/C_{ox}$, so that replacing this equation into Equation (4.5) one gets for the chemical potential the following expression [95].

$$\mu_{c\_unbiased} = \begin{cases} sign\left(\frac{C_{ox}}{q}V_{bias} + n_{e\_unbiased} - n_{h\_unbiased}\right)\hbar\nu_f \\ \quad \times\sqrt{\left|\frac{C_{ox}}{q}V_{bias} + n_{e\_unbiased} - n_{h\_unbiased}\right|\pi} & (4.6a) \\ \frac{\pi\hbar^2}{2m^*}\left(\frac{C_{ox}}{q}V_{bias} + n_{e\_unbiased} - n_{h\_unbiased}\right) & (4.6b) \end{cases}$$

The applied field experience under beam deflection also influences the scattering rate, Γ, which also takes a different value in this configuration of the beam due to the remote polar phonon scattering from the bottom dielectric. Sharma et al. [91] point out, however, that when the graphene is highly doped, the contribution due to remote polar phonon scattering rate is usually small [96] and can be ignored.

The performance of the graphene-based RF NEMS switch was modeled [91] after the experimental realization of Kim et al. [97], where the graphene beam had a length $L = 20$ μm and a width $w = 30$ μm and was suspended at a distance $g_0 = 300$ nm over the CPW signal line, which had a width $W = 15$ mm, and HfO_2 with a dielectric constant and a loss tangent $\varepsilon_r = 25$, $\tan\delta = 0.0098$ as the dielectric deposited over the signal line, with a thickness $t_d = 20$ nm; a thin dielectric leads to a higher ON/OFF capacitance ratio. The motivation for choosing a high-permittivity dielectric is pointed out by Sharma et al. [91] to be twofold, namely it reduces impurity scattering in graphene and it leads to better switch performance at high frequencies. The substrate utilized was high resistivity silicon ($\rho = 10\ Kohm - cm$). In the electromagnetic model, perfect electrical conductor (PEC) boundary conditions were assigned to the ground and signal lines. The model was summarized by the parameters in Table 4.1.

The microwave performance of the switch, captured by the calculation of its scattering parameters, was obtained in the frequency range of 1 GHz to 60 GHz. The insertion loss was 0.01–0.3 dB and 0.01–0.2 dB for monolayer and multilayer graphene-based switches, respectively, and the isolation was >10 dB for monolayer and >20 dB for multilayer graphene switches. It was surmised that the multilayer graphene switch exhibited a greater isolation than the monolayer one due to its lower sheet resistance which resulted in lower switch losses.

Table 4.1 Parameters derived for the model

Name	Monolayer	Multilayer	Reference
$1/\sigma$ $(\Omega/square)$	125	30	[98]
Doping	p-doped	p-doped	[98]
$n_{h\_undeflected}$ (cm^{-2})	9.43×10^{12}	4×10^{13}	[98]
$n_{e\_undeflected}$ (cm^{-2})	0	0	[98]
$\mu_{c\_undeflected}$ (eV)	0.365	0.92	(4.3)
$V_{pull\_in}(V)$	−0.3	−1.4	
$\mu_{c\_deflected}$ (eV)	0.451	0.344	(4.5)
τ (ps)	0.186	0.309	[98]

Source: Ref. [91].

4.3 MEMS/NEMS Varactors

RF MEMS/NEMS varactors are capacitors whose value may be controlled via the modification of one of its parameters, namely its area, its dielectric constant, or the distance separating its plates, upon the application of a voltage [83]. Varactors are typically utilized in electronics and radio frequency systems such as voltage-controlled oscillators (VCOs) and reconfigurable and tunable circuits.

The fundamental motivation behind the development of MEMS and NEMS varactors is their potential for exhibiting higher quality factors (Q) than varactors embedded/integrated in a semiconductor, due to the possibility of employing manufacturing processes that *decouple* their construction from the lossy substrate [10] and, thus, can operate at higher frequencies due to attaining higher self-resonance frequencies, for example, tens of GHz as opposed to a few GHz. In addition, since MEMS/NEMS varactors have no *pn*-junctions, they withstand large voltage swings without eliciting their nonlinearity, while displaying symmetric capacitance–voltage (C–V) characteristics [10]. One key aspects of this characteristic is the tuning range, that is, the range of capacitance variation that may be achieved before pull-in occurs in varactors whose inter-plate distance is varied electrostatically.

4.3.1 Nanoelectromechanical Varactors

4.3.1.1 Dual-gap MEMS/NEMS varactors

A prototypical parallel-plate RF MEMS varactor is shown in Figure 4.14.

In this design, to avoid pull-in, and thus extend the tuning range, use of separate control and signal electrodes (so-called dual-gap approach) with respective electrode to top plate separations d_2 and d_1 obeying the relationship $d_1 \leq d_2/3$ was employed. This resulted in the tuning range (TR)

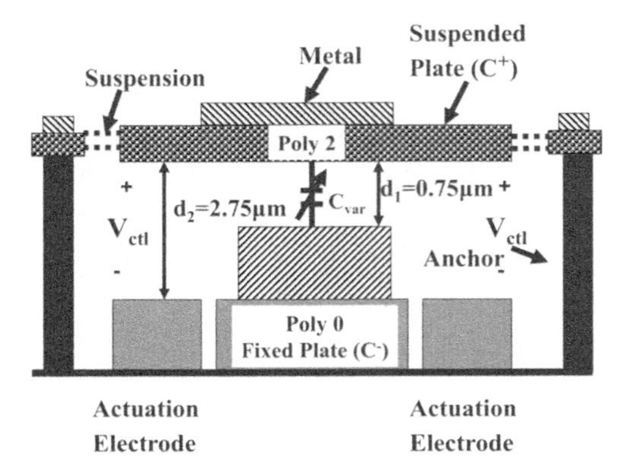

Figure 4.14 Cross-sectional view of MEMS varactor. V_{ctl} denotes the actuation or control voltage. Notice that the varactor is realized on top of (decoupled from) the substrate, which may be a silicon wafer but does not lower the varactor Q [82].

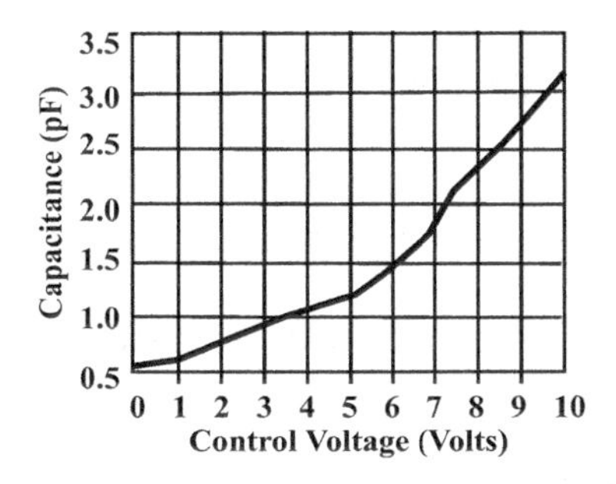

Figure 4.15 Measured capacitance characteristics of the varactor.

Source: Ref. [82].

being extended from the theoretical pull-in-limited value of 50% to more than 400% (Figure 4.15) [99]. An advantage of electrostatic actuation over other schemes is that it enables extremely low power consumption, in which power is consumed only during (switching) motion (compared to a digital inverter stage). In addition, using electrostatic actuation involves a relatively simple fabrication technology, a high degree of compatibility with a standard IC processes, ease of integration with planar and microstrip transmission lines, and a fast response (usually in the microsecond range). On the other hand, the main drawback of electrostatic actuation is the high actuation voltage (in the range 12–60 V) also associated with the large-gap actuator required in RF MEMS switches. Whenever the available supply voltage is limited, for example, to 3–5 V as in handheld phones, on-chip, high-voltage generators, such as the Dickson-type dc voltage multiplier circuit, may be incorporated [82].

4.3.1.2 MEMS/NEMS varactors via new materials

As indicated previously, the drawbacks of traditional RF MEMS varactors include high actuation voltages and relatively slow tuning speeds. The advent of new nanomaterials such as films made up of networks of single-wall carbon nanotubes (SWCNTs) has been proposed as a new direction for attaining superior performance NEMS varactors for mmWave and THz applications [100]. SWCNT-based networks were found to possess much lower Young's modulus, for example, from 60 MPa to 10 GPa, as opposed to 1 TPa. In these films, the tubes are connected to each other via van der Waals forces, and this determines the overall elasticity of the film. Accordingly, the Young's modulus of SWCNT films, as compared to films of traditional materials, exhibit lower actuation voltages [100]. The implementation of NEMS varactors utilizing SWCNT network-based membranes was first demonstrated by Generalov et al. [100]. A sketch of this design is shown in Figure 4.16.

Here, trenches are made in a wafer, at the bottom of which Au electrodes are formed, and the trenches are covered with the SWCNT film, thus forming a capacitance between the Au electrodes and the suspended SWCNTs film. The wafer employed was a high resistivity ($\rho = 10\ Kohm - cm$) silicon wafer; as mentioned previously, high resistivity wafers enable lower signal loss at high frequencies.

The fabrication of the device was as follows [100]. A 500-nm layer of Si_3N_4 was first deposited on the wafer; this acted as insulator between the SWCNT film and the wafer. Then, the trenches were patterned using a

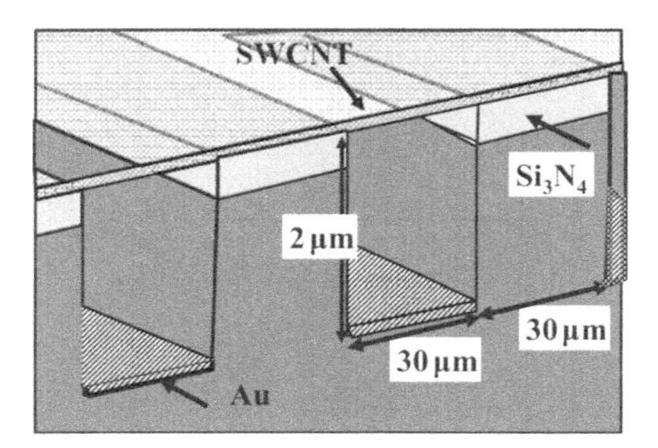

Figure 4.16 Design of the SWCNT MEMS varactor.

Source: Ref. [100].

Microtech LW 405 laser writer lithography on a 1.4-μm thick photoresist AZ 5214E. This was followed by the application of reactive-ion etching to form the 2-μm deep trenches. Next, after the etching, a 24-nm thick Au film was evaporated on the sample without removing photoresist, and this was followed by a lift-off step. This Au film formed the actuation electrodes. The SWCNT film was finally deposited on the top of the structure with a direct transfer method from the nitrocellulose filter after dry transfer. Some nanotubes from the SWCNT film can form a short circuit between the contacts. To preclude the possibility of some nanotubes falling into the trenches, Generalov et al. [100] recommended filling them with nanocellulose aerogel. This is a soft and flexible porous material that prevents short circuits between the contacts/actuation electrodes and the deflected SWCNT film. Since the nanocellulose aerogel has a porosity of 98%, its impact on the device is mainly mechanical (the SWCNT film pushes against it when it deflects) as its dielectric constant is close to that of air and, thus, close to unity [100]. As indicated in Figure 4.16, the dimensions of the trenches were 30 μm wide by 2 μm deep and had a length of 4 mm. The complete device contained 32 trenches, of which 16 were covered by the SWCNT film, and occupied an area of 2 mm × 2 mm [100].

Upon the application of the actuation voltage, the film displacement is as sketched in Figure 4.17, and the capacitance versus applied voltage is as shown in Figure 4.18.

It is seen that the capacitance deviation over the tuning voltage range is less than that without cellulose (Figure 4.18). This was determined to be due to the higher Young's modulus in the system when cellulose is filling the trench, namely 140 MPa, than when only the SWCNT film is present, whose Young's modulus is 80 MPa [100]. The varactor is tunable 100% with the measured capacitance varying from 3 nF to 5.8 nF over an applied voltage range of 10 V.

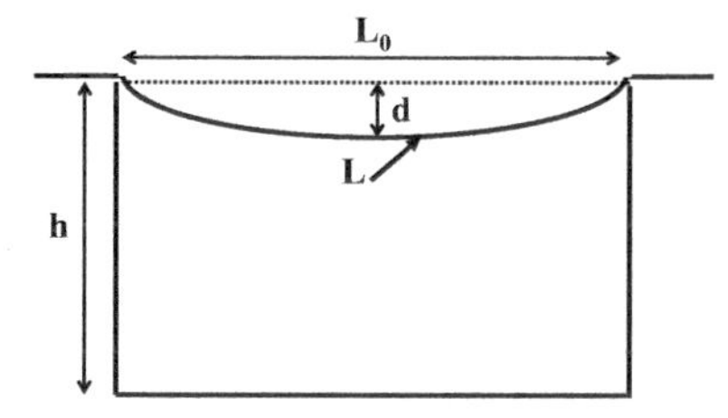

Figure 4.17 Cross section of the bent SWCNT film.

Source: Ref. [100].

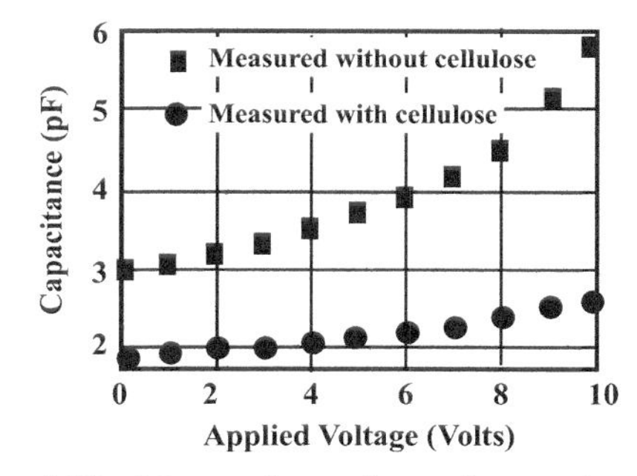

Figure 4.18 Measured capacitance for tested samples.

Source: Ref. [100].

4.4 MEMS/NEMS Resonators

RF MEMS/NEMS resonators are employed to provide the frequency-determining passive element in oscillators and filters. At lower frequencies, they may take the form of quartz crystals and MEM beams driven at their mechanical resonance frequency, whereas at higher frequencies they usually take the form of microwave cavities [10, 80, 101]. The advent of nanomaterials such as carbon nanotubes, graphene, and molybdenum disulfide has opened up the direction for attaining higher resonance frequencies with *nanomechanical* resonators. We begin our treatment of resonators with the traditional MEMS resonator type.

4.4.1 Nanoelectromechanical Resonators

4.4.1.1 Clamp–clamp RF MEMS resonators

The main parameters that capture the performance of MEMS/NEMS resonators are their resonance frequency and quality factor (Q) [10, 102]. The resonance frequency, as we saw previously, is a function of the beam geometry, stiffness, and material density. The Q is a function of the loss mechanisms, in particular, energy loss to the substrate through the beam anchor.

An approach to address this latter issue of energy loss is the clamp–clamp MEMS beam resonator shown in Figure 4.19 [102]. In this design, the input voltage, Vi, excites the beam into vibrating laterally, thereby producing an output current, i_O, which is a function of both the input voltage and a DC bias voltage, Vp. In particular, in this so-called free–free (FF) beam resonator, the Q is maximized because in it the anchor dissipation was reduced by suspending it using quarter wavelength flexural support beams. This may be

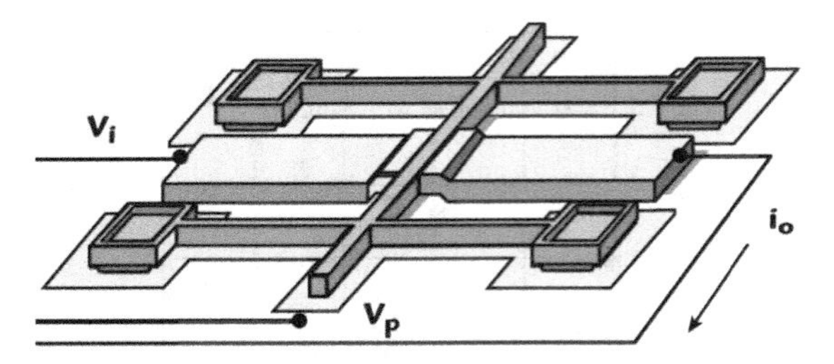

Figure 4.19 Two-port lateral free-free micromechanical resonator.

Source: Ref. [102]. Reprinted with permission from *Microwave Journal*.

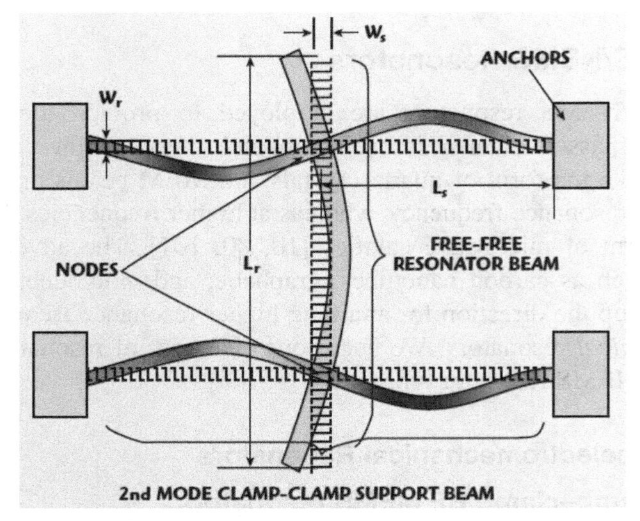

Figure 4.20 Simulated displacement contour plot of a lateral FF-beam micromechanical resonator.

Source: Ref. [102]. Reprinted with permission from *Microwave Journal*.

appreciated from Figure 4.20, which shows (ANSYS Multiphysics simulations) that the junctions between the beam and the suspensions are nodes of the beam vibration deformation, that is, they experience, ideally, zero motion.

The resonance frequency of the FF beam is given by [102],

$$f_0 = 1.03\sqrt{\frac{E}{\rho}\frac{W_r}{L_r^2}} \tag{4.7}$$

where E is beam's Young's modulus, ρ is its density, L_r is its length and W_r is its width. The key to achieving zero displacement at the beam–suspension junction is that by having the support beams resonate in their second mode, relative to the fundamental frequency mode of the FF beam, a high impedance is located at the junctions through which very little energy is transferred, and hence, very little energy is dissipated. To set the support beams to vibrate in the second mode, their length has to be set to [102],

$$L_S = 1.683 \left(\sqrt{\frac{E}{\rho}} \frac{W_S}{f_0} \right)^{\frac{1}{2}} \tag{4.8}$$

where L_S is the length of the support beam and W_s is its width. A scanning electron micrograph of the FF resonator and its measured performance are shown in Figures 4.21 and 4.22. The low amplitude of the spectrum is caused by the impedance mismatch between the motional resistance and the 50 Ω termination provided by the network analyzer.

4.4.1.2 MEMS/NEMS resonators via new materials

NEMS resonators may be produced either via "top–down" methods (fabrication processes involving material deposition, photolithography and chemical etching, as in the example discussed above) or "bottom–up" methods (processes producing intrinsically small objects such as carbon nanotubes and graphene sheets). The advantage of NEMS resonators obtained from

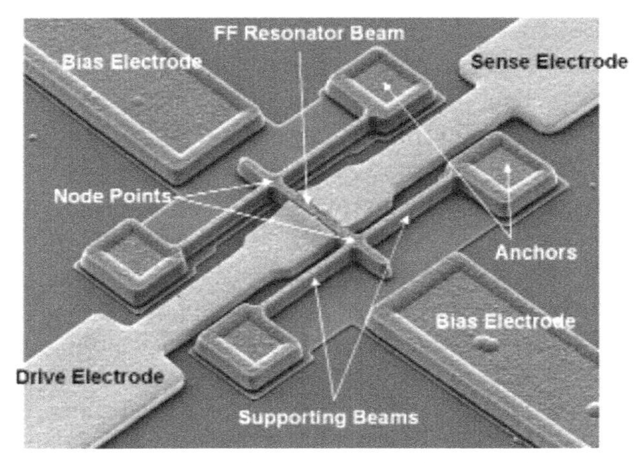

Figure 4.21 SEM photograph of the fabricated lateral FF-beam micromechanical resonator.
Source: Ref. [102]. Reprinted with permission from *Microwave Journal*.

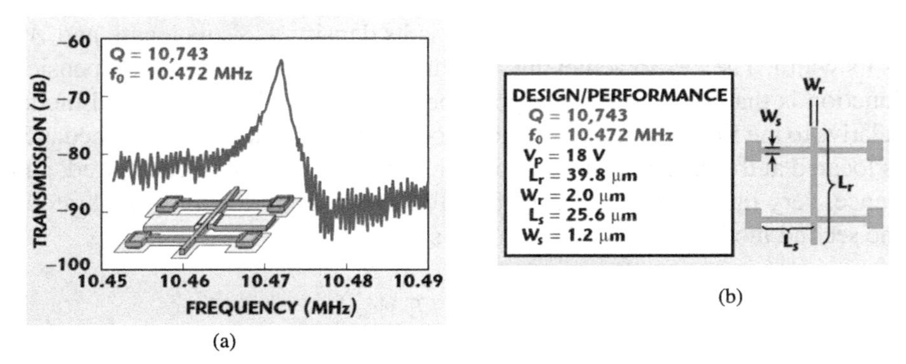

Figure 4.22 Measured frequency spectrum of the lateral FF-beam micromechanical resonator.

Source: Ref. [102]. Reprinted with permission from *Microwave Journal*.

"bottom–up" fabrication procedures, however, is that their surfaces are natural planes and, thus, lack defects that can induce energy losses, thereby exhibiting higher Qs [103]. In resonators for application as frequency-determining elements in, for example, oscillators, the desirability of high Q is due to its inverse relation to the oscillator phase noise, that is, the higher the Q, the lower the phase noise, so the more stable the output frequency produced. On the other hand, in the context of nanoresonators which are applied to sensing extremely low masses, the sensitivity of the nanoresonator is inversely proportional to its Q factor, and a number of additional loss mechanisms that become important at nanoscales need to be suppressed. These energy loss mechanisms, in addition to external anchor energy loss, are said to include intrinsic nonlinear scattering mechanisms, the effective strain mechanism, edge effects, grain boundary-mediated scattering losses, and the adsorbate migration effect [105–107]. This topic of determining noise in NEMS is currently one of much research, as the definitive causes of these loss mechanisms seem elusive and are not settled knowledge yet [106]. Nevertheless, we next give examples of NEMS resonators implemented in graphene and MoS_2.

Investigation of the properties of graphene applied to resonators appears to have been initiated by Bunch et al. [104], and a prototypical example of a NEMS resonator based on graphene sheets was advanced by him and his co-workers, where various properties of these resonators, including resonance frequency, spring constant, built-in tension, and quality factor, were studied [104].

One of the key features of graphene, utilized in conjunction with these mechanical properties, is its outstanding electronic properties such as its high electron mobility that enables ultrafast operation, for example, THz field-effect transistors (FETs) and radiation detectors [108]. High mobility also facilitates resonant plasma wave excitation in those structures which, when configured as micron-length graphene plasma resonators, make them compatible with the resonance frequencies of plasma oscillations in the THz regime [108]. Plasma oscillations are wave-like propagation of the electron density of a medium [72].

Recently, the utilization of graphene sheets as the gate of a field-effect transistor-like device for detecting THz waves was reported by Svintsov et al. [108] (Figure 4.23) in analogy to the resonant-gate transistor introduced more than 50 years ago [109]. The device structure consisted of an FET in which the gate was a floating graphene beam and the channel was a graphene sheet (Figure 4.23a). To detect the incoming THz signal, the amplitude-modulated

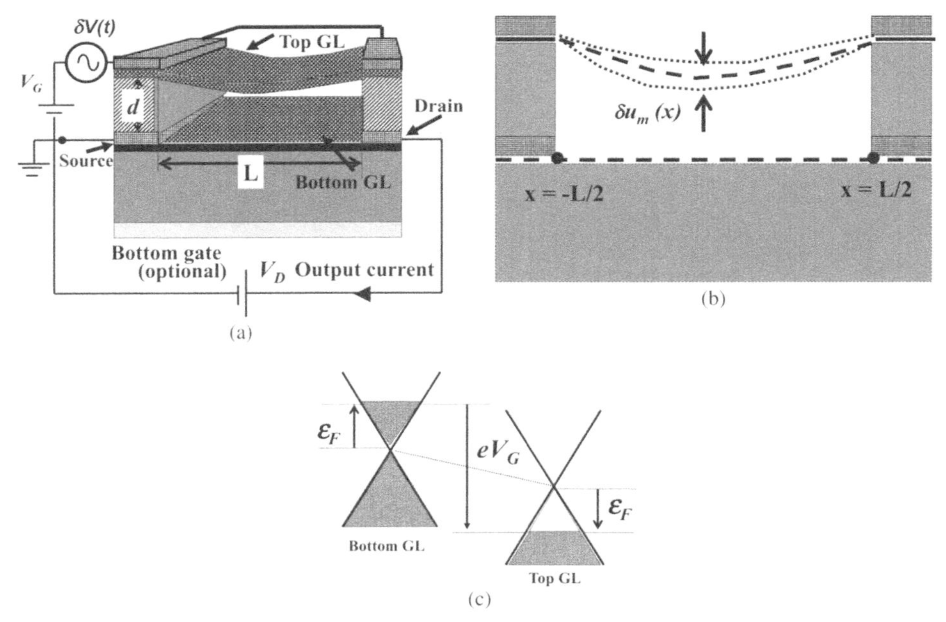

Figure 4.23 (a) Sketch of the proposed detector structure, (b) schematic view of suspended GL static deflection (bold line) and its swinging amplitude $\delta u_m(x)$ due to the driving GHz force (thin lines), and (c) band diagram of the double-GL structure under applied voltage V_G (filled areas are occupied by electrons).

Source: Ref. [104].

signal $\delta V(t)$ at a modulation frequency ω_m is collected as the THz antenna output and applied between the top and bottom layers. This modulation signal sets the graphene beam into mechanical oscillation at a frequency close to that modulating the THz carrier, which is in the range of GHz. On the other hand, the THz carrier frequency excites the plasma oscillations in the graphene; this has been determined to be responsible for an increase in the electric field between the top and bottom graphene layers (GLs) [108]. A detailed analysis of the device was carried out by Svintsov et al. [108] in two steps, namely first dealing with the plasma response, and then with the mechanical response.

The plasma response is determined by formulating the dynamics of the electron density following the application of the driving electric field derived from the modulated THz signal applied between the top and bottom GLs (Figure 4.23a). The GL layers have length and width L and $W \gg L$, respectively, and are separated by a vertical distance d. The GLs are biased by the DC voltage V_G, applied between the top and bottom GL, which induces an electron density in the top GL and a hole density in the bottom GL, of equal magnitude ρ, as represented in Figure 4.23c, and are given by,

$$CV_G = e\rho \tag{4.9}$$

where

$$C = \frac{C_G C_q/2}{C_G + C_q/2} \tag{4.10}$$

with $C_G = \varepsilon_0/d$ being the gate capacitance per unit area and $C_q = e^2\partial\rho/\partial\varepsilon_F$ being the *quantum capacitance*, which represents the variation of the Fermi energy with electric field strength between the GLs. Svintsov et al. [108] indicate that, for distances between the GLs $d \geq 20$ nm and at room temperature or for high carrier densities ρ, $C_q \gg C_G$ and $C \approx C_q \cdot V_D$ applied across the bottom GL, which plays the role of channel of an FET, induces a DC current through it.

Now, the signal to be detected, proceeding from the antenna, is represented by,

$$\delta V(t) = \delta V_m \cos(\omega t)[1 + m\cos(\omega_m t)] \tag{4.11}$$

where δV_m is its amplitude, ω is the THz carrier frequency, m is the amplitude modulation depth, and ω_m is the modulation frequency. This signal, which may be represented by the small-signal voltage $\delta V e^{-i\omega t}$ applied between the top and bottom GLs, elicits a plasma wave response that manifests itself as a potential difference $\delta\varphi_+ - \delta\varphi_-$ between the top and bottom GLs corresponding to the charge density perturbation $e\delta\rho$ concomitant with the plasma wave given by,

$$C(\delta\varphi_+ - \delta\varphi_-) = -e\delta\rho \tag{4.12}$$

By charge continuity, $\partial(e\rho)/\partial t = -\partial J/\partial x$, with current density given by Ohm's law, $J = \sigma E = -\sigma\partial\varphi/\partial x$, the charge perturbation in the top and bottom GLs is given, respectively, by,

$$i\omega e\delta\rho = \sigma_+ \frac{\partial^2 \delta\varphi_+}{\partial x^2} \rightarrow i\omega e\delta\rho \frac{1}{\sigma_+} = \frac{\partial^2 \delta\varphi_+}{\partial x^2} \tag{4.13}$$

and

$$i\omega e\delta\rho = -\sigma_- \frac{\partial^2 \delta\varphi_-}{\partial x^2} \rightarrow i\omega e\delta\rho \frac{1}{\sigma_-} = \frac{\partial^2 \delta\varphi_-}{\partial x^2} \tag{4.14}$$

where σ_+ and σ_- are the sheet conductivities of the top and bottom GLs. Adding Equations (4.13) and (4.14), we obtain,

$$i\omega e\delta\rho \left(\frac{1}{\sigma_+} + \frac{1}{\sigma_-}\right) = \frac{\partial^2}{\partial x^2}(\delta\varphi_+ - \delta\varphi_-) \tag{4.15}$$

and using Equation (4.11) in Equation (4.15), we have,

$$\frac{\partial^2}{\partial x^2}(\delta\varphi_+ - \delta\varphi_-) = -i\omega eC \left(\frac{1}{\sigma_+} + \frac{1}{\sigma_-}\right)(\delta\varphi_+ - \delta\varphi_-) \tag{4.16}$$

or

$$\frac{\partial^2}{\partial x^2}(\delta\varphi_+ - \delta\varphi_-) + i\omega eC \left(\frac{1}{\sigma_+} + \frac{1}{\sigma_-}\right)(\delta\varphi_+ - \delta\varphi_-) = 0 \tag{4.17}$$

which may be written as,

$$\frac{\partial^2}{\partial x^2}(\delta\varphi_+ - \delta\varphi_-) + \gamma_w^2(\delta\varphi_+ - \delta\varphi_-) = 0 \tag{4.18}$$

which is a wave equation with propagation constant, $\gamma_\omega^2 = i\omega eC\left(\frac{1}{\sigma_+} + \frac{1}{\sigma_-}\right)$. The plasma wave embodied by the above equation was envisioned as propagating in a circuit as shown in Figure 4.24.

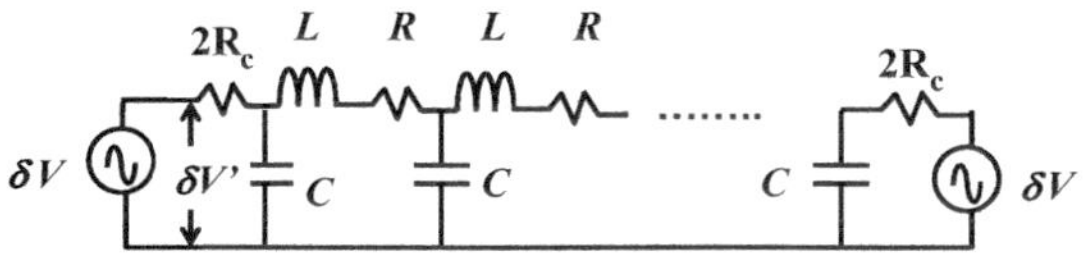

Figure 4.24 Transmission line equivalent of the double-graphene layer structure, including contact resistance R_c, geometric inter-GL capacitance C, sheet inductance L, and sheet resistance R.

Source: Ref. [108].

The circuit elements in this transmission line model, assuming Drude's conductivity in the GLs, namely,

$$\sigma_{\pm} = \frac{e^2 \varepsilon_F / (\pi \hbar^2 \nu_{\pm})}{1 + i\omega / \nu_{\pm}} \tag{4.19}$$

where $\nu_{\pm}$ represents the carrier collision frequencies in their respective layers, are given by,

$$L = \frac{2\pi \hbar^2}{e^2 \varepsilon_F} \tag{4.20}$$

and

$$R = \frac{\pi \hbar^2 (\nu_+ + \nu_-)}{e^2 \varepsilon_F} \tag{4.21}$$

Defining the plasma wave propagation velocity in a gated graphene sheet, S, by,

$$S = \sqrt{\frac{e^2 \varepsilon_F}{\pi \hbar^2 C}} \tag{4.22}$$

the propagation constant may be expressed in a more transparent way as,

$$\gamma_m = \frac{\sqrt{\omega[2\omega + i(\nu_+ + \nu_-)]}}{S} \tag{4.23}$$

As usual, to find the solution of the differential equation in Equation (4.18), it is necessary to apply the pertinent boundary conditions. In this case, these are given by the voltages at the left and right ends of the channel, namely,

$$[\delta\varphi_+ - \delta\varphi_-]_{x=\pm L/2} = \frac{\delta V'}{2} \tag{4.24}$$

The prime in $\delta V'$ implies that the voltages driving the transmission line are different from δV due to a voltage drop across the contact resistance to the GLs. In particular, if δj is the incoming current density and R_C is the contact resistance, we have,

$$\delta V' = \delta V - 2\delta j R_C \tag{4.25}$$

With these boundary conditions, the solution to Equation (4.18) is found to be [108],

$$\delta\varphi_+ - \delta\varphi_- = \delta V h_\omega \cos(\gamma_\omega x) \tag{4.26}$$

where

$$h_\omega = \left[\cos\left(\frac{\omega_m L}{2}\right) - \frac{i\omega C R_C}{\gamma_m}\sin\left(\frac{\omega_m L}{2}\right)\right]^{-1} \tag{4.27}$$

denotes the dimensionless plasma resonant factor, and the first resonant frequency (assuming the collision frequency much smaller than the signal frequency), $\nu \ll \omega$, is given by,

$$\Omega = \frac{\pi S}{\sqrt{2}L} \tag{4.28}$$

To calibrate our intuition, for a gate voltage $V_G = 2$ V across a distance $d = 50$ nm, which, according to Svintsov et al. [108], corresponds to a plasma wave speed $S \approx 5 \times 10^6$ m/s, the concomitant first resonant frequency is $\Omega/2\pi = 1.8$ THz. As usual, in the neighborhood of the resonance, the resonance factor may be expressed by a Lorentzian curve, expressed as [108],

$$|h_\omega|^2 \approx \frac{0.8 Q_p^2}{1 + 4Q_p^2(\omega/\Omega - 1)^2} \tag{4.29}$$

where Q_p represents the quality factor of the plasma oscillations and is given by,

$$Q_p = \left[\frac{\nu_+ + \nu_-}{2\Omega} + \frac{4LCR_C\Omega}{\pi^2}\right]^{-1} \tag{4.30}$$

In Equation (4.28), the response is characterized by a magnitude and width determined by the average collision frequency, $(\nu_+ + \nu_-)/2$ and the contact resistance. Using a realistic value for the contact resistance of $R_C = 100\ Ohm - \mu m$, a plot of $|h_\omega|^2$ is shown in Figure 4.25.

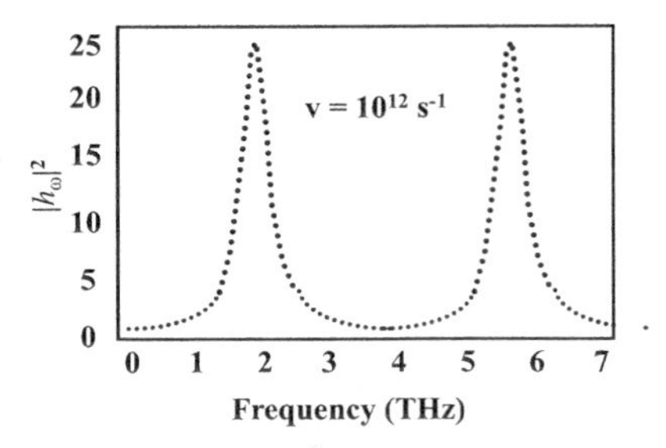

Figure 4.25 Plasma response function $|h_\omega|^2$ versus carrier frequency. The contact resistance is $R_C = 100$ *Ohm* $- \mu m$, and the collision frequency is $\nu = 10^{12}\ s^{-1}$. The structure length $L = 1$ μm, and the velocity of the plasma waves is $S \approx 5 \times 10^6\ m/s$. The Q-factor is $Qp \approx 25$.

Source: Ref. [108].

As seen previously, due to the applied gate bias, V_G, opposite charges are induced in the top and bottom GLs, and these opposite charges give rise to a nonlinear electrostatic force [18] between the GLs.

If, in the presence of this force, an additional oscillating force of frequency ω, such as that derived from the small signal being detected, is applied, a so-called ponderomotive force, quantified by [110],

$$F_p = -\frac{e^2}{4m\omega^2}\nabla(E^2) \tag{4.31}$$

is also experienced by the GLs. The average (the averaging time is taken as a period of time much less than $2\pi/\omega_m$) ponderomotive force of attraction between the GLs was given by Svintsov et al. [108] as,

$$\begin{aligned}\overline{F}_p(x,t) = {} & \frac{CV_G^2}{2d} + \frac{CV_G^2}{2d^2}\delta u(x,t) \\ & + |h_\omega|^2 |\cos(\gamma_\omega x)|^2 \frac{C\delta V_m^2}{4d} \\ & \times \left[\left(1 + \frac{m^2}{2}\right) + 2m\cos(\omega_m t) + \frac{m^2}{2}\cos(2\omega_m t)\right]\end{aligned} \tag{4.32}$$

This expression may be interpreted as follows [108]. The average force of attraction is the result of the constant gate voltage V_G, represented by the first term; the time-varying distance between the top and bottom GLs, represented by the second term; and rectification of the amplitude modulated signal, represented by the third term. This third term, in turn, contains a DC component, which may be used for detecting unmodulated signals, a term at the modulation frequency, ω_m, which induces forced oscillations of the top GL beam at this frequency, thus representing a mechanical response most pronounced when ω_m is near the beam resonance frequency and the double frequency component $2\omega_m$. When the double frequency component is well above the beam resonance frequency, this cannot follow the vibration, and its displacement at this frequency tends to be negligible [18]. The average force given above must elicit a displacement, the nature of which is dealt with subsequently.

In response to a force, the GL beam will displace a distance $u(x,t)$. The relationship between the force and the corresponding displacement $u(x,t)$ is arrived at by solving the elasticity equation for graphene [111]. This equation was expressed by Svintsov et al. [108] as,

$$\rho_{gr}\omega_m(\omega_m - i\upsilon_m)\delta u_m + T\frac{\partial^2 u_m}{\partial x^2} = -\frac{CV_G^2}{d^2}\delta u_m - \frac{m}{2}\frac{C\delta V_m^2}{d}|h_\omega|^2|\cos(\gamma_\omega x)|^2 \tag{4.33}$$

where $\rho_{gr} = 7 \times 10^{-7}$ kg m^{-2} is the mass density of graphene, m is the modulation index, ν_m represents the mechanical damping, $u(x,t) = u_0(x) + \delta u(x,t)$, with $u_0(x)$ representing the average displacement and $\delta u(x,t) = \delta u_m(x)e^{-i\omega_m t}$ representing the time-varying portion of the displacement, oscillating at the modulation frequency and T is the elastic force density given by,

$$T = Eh\delta x \tag{4.34}$$

where $Eh = 340\ Nm^{-1}$ is the two-dimensional elastic stiffness and *dx* is the tensile strain. In Equation (4.33), the first term on the right of the equal sign has been indicated to be responsible for reducing the resonance frequency of the beam [108] by an amount $\omega_{shift} = (V_G/d)\sqrt{C/\rho_{gr}}$, which for $V_G - 2V$ and $d = 50$ nm yield, $\omega_{shift}/2\pi \approx 100\ MHz$. Equation (4.32) may be further simplified by defining $\tilde{\omega}_m^2 = \omega_m(\omega_m - i\nu_m) + \omega_{shift}^2$ and the transverse velocity of sound as $c_s = \sqrt{T/\rho_{gr}}$ to yield,

$$c_S^2\frac{\partial^2 u_m}{\partial x^2} + \tilde{\omega}_m^2\delta u_m = -\frac{m}{2}\frac{C\delta V^2}{\rho_{gr}d}|h_\omega|^2|\cos(\gamma_\omega x)|^2 \tag{4.35}$$

For the clamped–clamped beam, in which there is zero displacement at $x = \pm L/2$ (Figure 4.23c), the x-average of the displacement was given as [108],

$$\langle \delta u_m(x) \rangle = \frac{m}{4} \frac{C\delta V^2}{\rho_{gr} d |\tilde{\omega}_m^2|} |h_\omega|^2 |H_{\omega,\omega_m}| \tag{4.36}$$

from which the gate-channel capacitance may be estimated. In Equation (4.36), the factor,

$$|H_{\omega,\omega_m}| = \frac{\tan \gamma_m L}{\gamma_m L} \left[\frac{\cos \gamma'_\omega L}{1 - (\gamma'_\omega / \gamma_m)} + 1 \right] - \left[\frac{\sin(\gamma'_\omega L)/\gamma'_\omega L}{1 - (\gamma'_\omega / \gamma_m)^2} + 1 \right] \tag{4.37}$$

where it is assumed again that $\nu \ll \Omega$, that is, the damping is much less than the frequency, $\gamma_m = \tilde{\omega}_m/(2c_S)$, $\gamma'_\omega = \mathrm{Re}(\gamma_\omega)$ and $\mathrm{Im}(\gamma_\omega) \approx 0$, denotes the *mechanical resonance factor* and captures the most pronounced resonances. Figure 4.26 shows a plot of $|H_{\omega,\omega_m}|$. The resonance may be described by a Lorentzian curve given by,

$$|H_{\Omega,\omega_m}| \approx \frac{1.1\, Q_m}{\sqrt{1 + 4\, Q_m^2 \left(1 - \frac{\sqrt{\omega_m^2 + \omega_{m\backslash shift}^2}}{\Omega_m}\right)^2}} \tag{4.38}$$

where $Q_m = \Omega_m / \nu_m$ is the quality factor of the mechanical resonator which for graphene resonators tuned to the frequencies of hundreds of MHz are of the order of 100.

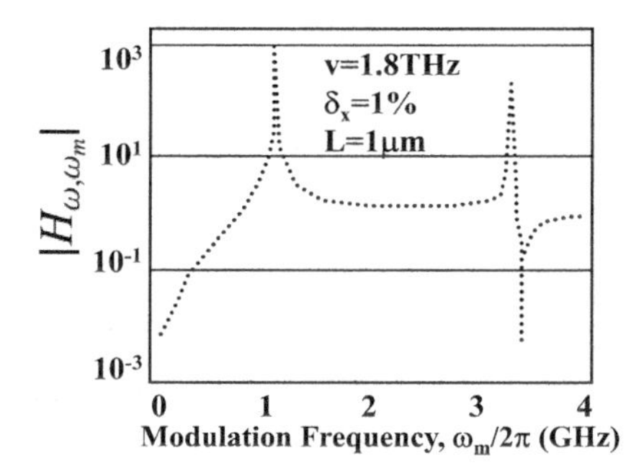

Figure 4.26 Mechanical response function $|H_{\omega,\omega_m}|$ versus modulation frequency ω_m. Structure length $L = 1$ μm, tensile strain of top GL is $\delta_x = 0.01$, quality factor $Q_m = 10^3$.

Source: Ref. [108].

4.5 Summary

In this chapter, we have dealt with the fundamental physics of MEMS/NEMS devices, in particular, switches, varactors, and resonators. In our treatment, we have presented miniaturized devices fabricated via the "top–down" approach, in which material deposition, photolithography, and chemical etching are some of the key steps to fabricate the devices. In an almost parallel fashion, we have also presented versions of these devices fabricated via "bottom–up" methods. The emphasis, however, has not been on the fabrication of the devices but on their structures and device physics.

5

Understanding MEMS/NEMS for Energy Harvesting

5.1 Introduction

The fundamental motivation for energy harvesting, in the context of the Internet of Things (IoT), is energy autonomy. That is, since wireless sensor nodes are isolated and remote (inaccessible), they need to operate with limited battery power or capture (harvest) energy from the environment.

A number of approaches are employed to harvest energy from the environment, including capturing ambient energy (such as radio signals from TV stations, cell phone base stations, etc.) for converting it to DC power; converting solar energy into DC power; converting thermal gradients into energy; and converting environmental mechanical vibrations into energy [112]. The prototypical architecture for an IoT node is depicted in Figure 5.1 [113].

In this chapter, we review various approaches to energy harvesting, in the context of the IoT, including, whenever possible, their implementation as MEMS/NEMS devices.

5.2 Wireless Energy Harvesting

Since the atmospheric environment is permeated by radio frequency signals from a wide variety of sources, for instance, TV broadcast stations and cell phone base stations, it is natural to explore the possibility of capturing these signals to derive from them DC power for IoT devices. Such a feasibility study was carried out by Kitazawa et al. [114]. In their study, they calculated and measured the power spectral density of the power received in a common suburban area, as derived from ambient signals (Figure 5.2). In particular, for an antenna of gain Gt and transmitting power Pt, the power flux density

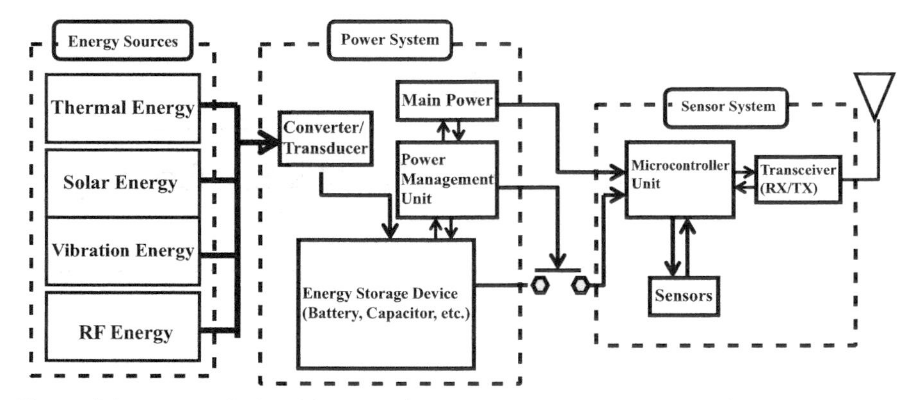

Figure 5.1 Prototypical architecture of an energy-harvesting-enabled wireless sensor platform.

Source: Ref. [113].

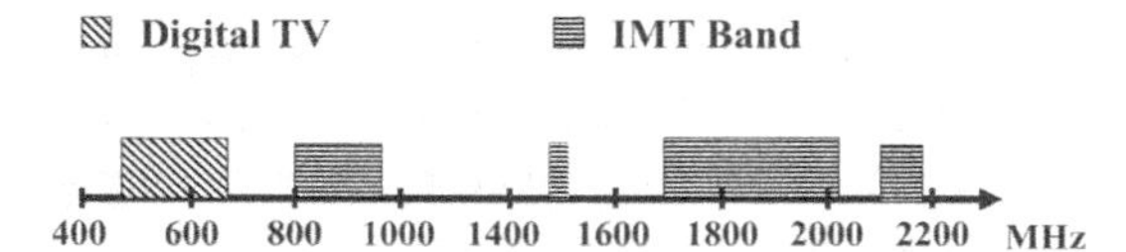

Figure 5.2 TV and mobile telephone frequency allocations chart.

Source: Ref. [114].

(PFD) received at a distance R is given by,

$$S = \frac{P_t G_t}{4\pi R^2} K \tag{5.1}$$

where K = 2.56 is a factor representing the reflection coefficient [114].

They found that the median value of received power from the 800 MHz band from cell phone base stations, namely −25 dBm, was, on average, 13 dBm higher than that from Digital TV broadcasting.

On the other hand, while one would, logically, ask whether the radio frequency (RF) power in residential and suburban areas coming from AM and FM radio, TV broadcasting, cell phones, and wireless local area networks (LAN) should also be targeted for harvesting, these had previously been determined to be much weaker than that from Digital TV and base stations (BS), and so, their pursuit was abandoned. Table 5.1 represents the specifications for the power transmitted by Digital TV and base stations, and Figure 5.3 shows plots of the respective calculated received power flux density versus distance.

Table 5.1 Transmission power specifications

Name	Digital TV	Cell Phone Base Station
Transmission power	3 kW	3 W
Effective radiated power	25 kW	500 W

Source: Ref. [114].

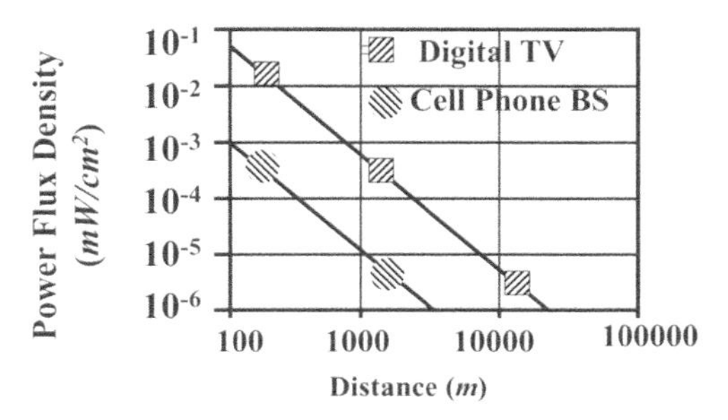

Figure 5.3 Calculated power flux density as a function of distance.

Source: Ref. [114].

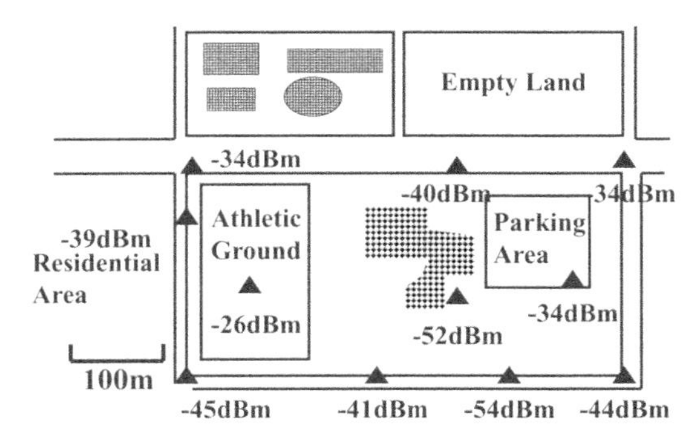

Figure 5.4 Measured results of ambient RF signals: Digital TV. Fill in graphic denotes buildings.

Source: Ref. [114].

The results of the measurements alluded to above in a suburban area are shown in Figures 5.4 and 5.5.

The measurement conditions were as follows. Using a standard dipole antenna and a spectrum analyzer, the TV transmission was measured in an area approximately 10 km away from the TV station.

On the other hand, the 800 MHz signal emitted by a cell phone base station located at the top of a building, at an altitude of 50 m in the same area, was measured. The received signal was not line of sight because the building in question was in a depressed terrain, as a result of which the measured

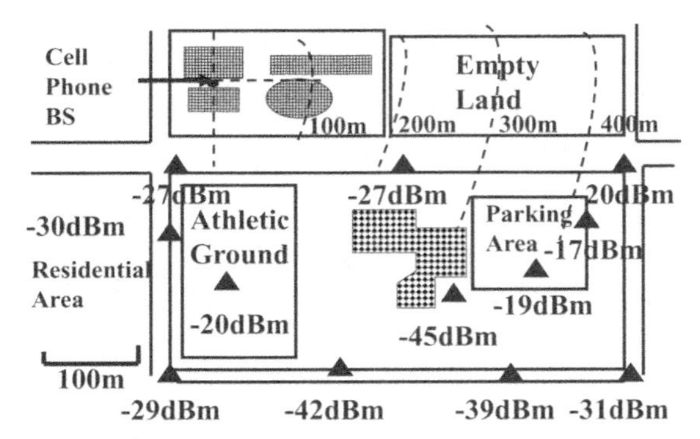

Figure 5.5 Measured results of ambient RF signals: Cell phone base station. Fill in graphic denotes buildings.

Source: Ref. [114].

Table 5.2 Characteristics of ambient RF energy sources

Characteristic	Ambient RF Energy
Power density	0.0002–1 μW/cm^2
Output	3–4 V (open circuit)
Availability	Continuously
Weight	2–3 g
Pros	• Antenna can be integrated into frame • Widely available
Cons	• Distance-dependent • Is a function of available power source

received power was lower than that predicted by Equation (5.1). In particular, received power lower than −30 dBm was observed in the range of 500 m from the cell phone BS. The general characteristics of ambient RF power were discussed in Ref. [114] and are summarized in Table 5.2.

Once received, the ambient RF power must be converted to DC power. This is addressed in the subsequent section.

5.2.1 RF-DC Conversion Circuit

The problem of RF-to-DC conversion is the familiar one of AC rectification, except that here we are dealing with signals at frequencies in the hundreds and thousands of MHz. A prototypical RF ambient energy architecture is shown in Figure 5.6.

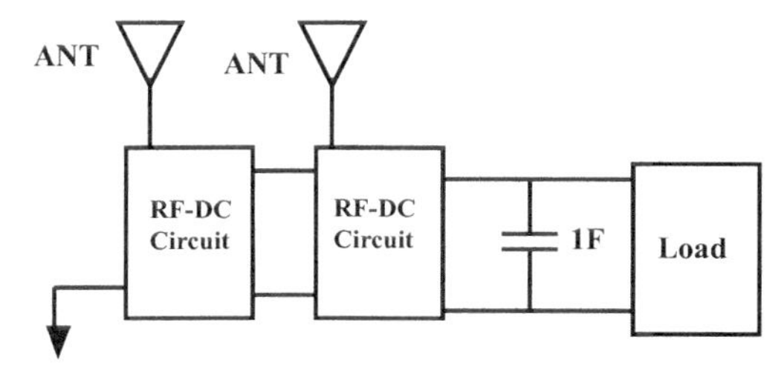

Figure 5.6 Architecture of prototypical energy harvesting system. The 1F capacitor is for energy storage.

Source: Ref. [114].

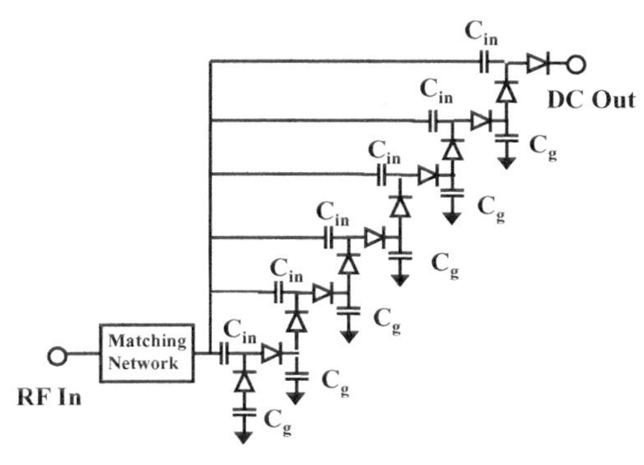

Figure 5.7 Schematic of a 6-stage RF-DC conversion circuit.

Source: Ref. [114].

An experimental realization of the RF-DC circuit was implemented and tested by Kitazawa et al. [114] (Figure 5.7). In this circuit, the cascade connection of single-stage RF-DC conversion circuits is such that the output voltage of the 1st RF-DC conversion circuit is added to the first diode of the 2nd RF-DC conversion circuit and so on. With capacitors C_{in} of 10 pF and C_g of 1 pF, and a Hitachi Semiconductor HSB276AS Schottky barrier diode, a 1 GHz – 20 dBm signal applied at the input of this 6-stage circuit yielded an output voltage of 20 mV, which was about over 10 times higher than that obtained with a single-stage diode rectifier.

The performance of this circuit is given in Table 5.3.

The development of the storage energy, that is, the charging of the storage capacitor, is not instantaneous, but takes some time. For the circuit in question (Figure 5.7), the measured accumulated (charging) voltage versus time on a 1F capacitor was as depicted in Figure 5.8 [114].

Table 5.3 RF-to-DC converter efficiency

Input Power (dBm)	Open Circuit Voltage (mV)	Terminal Voltage@ $R_{Load} = 10\ k\Omega$ (mV)	Conversion Efficiency (%)
5	5590	790	19.7
10	3240	307	9.4
15	1729	97.8	3.0
20	1816	27.4	0.8

Source: Ref. [114].

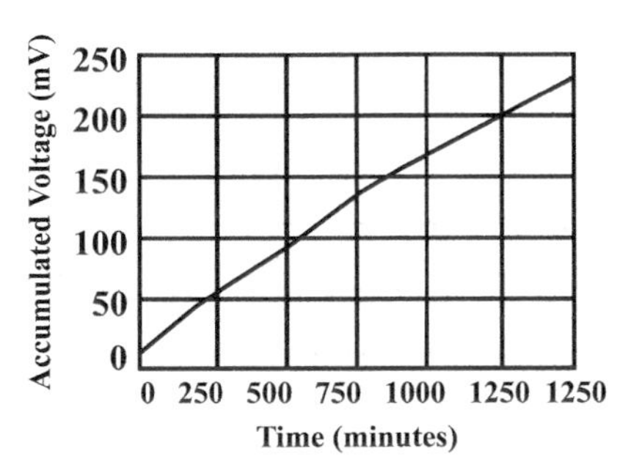

Figure 5.8 Accumulated voltage by RF energy harvesting.

Source: Ref. [114].

In general, the sources of RF ambient energy possess different frequencies due to the adoption of different standards around the world. Information pertinent to several countries is provided in Figure 5.9 and Table 5.4. Piñuela et al. [115] have reported wireless energy harvesting circuit architectures for the collection of multiple frequencies, thus enabling reception and power extraction from many portions of the available RF spectrum.

5.2.2 Resonant Amplification of Extremely Small Signals

As shown in Figure 5.1, the prototypical architecture of an energy-harvesting-enabled wireless sensor platform has three stages, namely the energy sources, the power system, and the sensor system, which includes the transceiver interfacing with the sensors. To conserve energy, IoT wireless nodes require low-power RF transceivers that remain dormant while not in use and that can be woken-up by extremely low-power signals [116].

One of the major obstacles for a transceiver is waking-up upon detection of low-power signals possessing an amplitude significantly below the threshold voltages of state-of-the-art semiconductors or alternative MEMS devices.

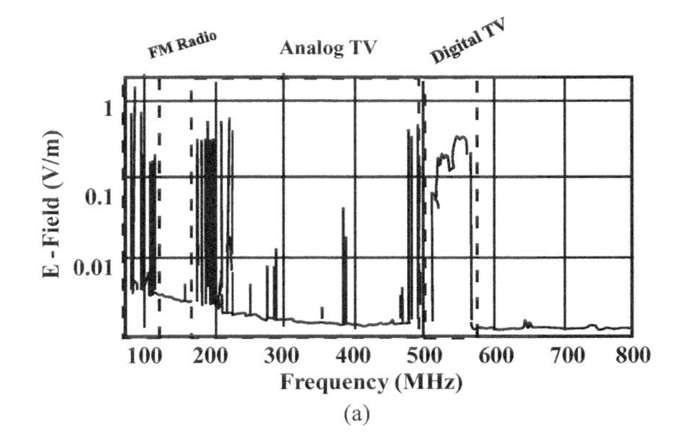

(a)

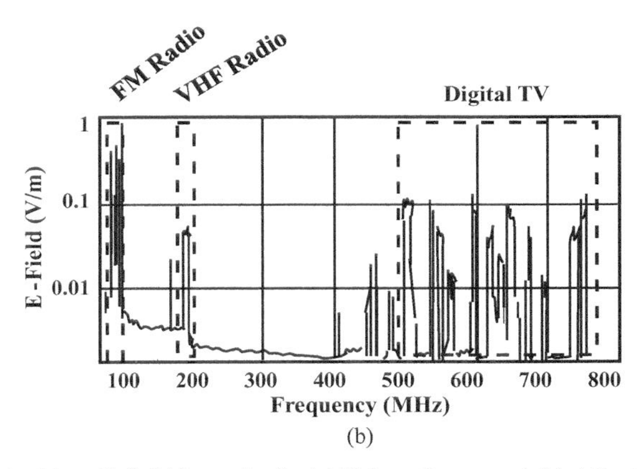

(b)

Figure 5.9 Ambient E-field intensity in (a) Tokyo, Japan and (b) Atlanta, GA, USA. *Source*: Ref. [113].

Another issue is incorporating high-frequency selectivity to avoid false wake-up caused by interference signals in the adjacent RF spectrum. These issues have been addressed by Lu et al. [116] in their piezoelectric RF resonant voltage amplifier concept.

Figure 5.10 depicts the approach of Lu et al. [116] to detecting or amplifying faint signals. The signal received by the antenna is coupled to a piezoelectric voltage amplifier and then subsequently to the rectifier and from this to a comparator to decide whether or not to wake up the transceiver.

The key element of the approach is the piezoelectric voltage amplifier [117]. In this *passive* (no DC bias) amplifier, amplification is effected by virtue of the generation of an acoustic wave that, via the piezoelectric effect, is accompanied by an electric field. The device gain is proportional to the

Table 5.4 RF power density in London, UK

Band	Frequency (MHz)	Average S_{BA} (nW/cm^2)	Maximum S_{BA} (nW/cm^2)
DTV	470–610	0.89	460
GSM 900 (MTx)	880–915	0.45	39
GSM 900 (BTx)	925–960	36	1930
GSM 1800 (MTx)	1710–1785	0.5	20
GSM 1800 (MTx)	1805–1880	84	6390
3G (MTx)	1920–1980	0.46	66
3G (BTx)	2110–2170	12	240
WiFi	2400–2500	0.18	6

Source: Ref. [114].

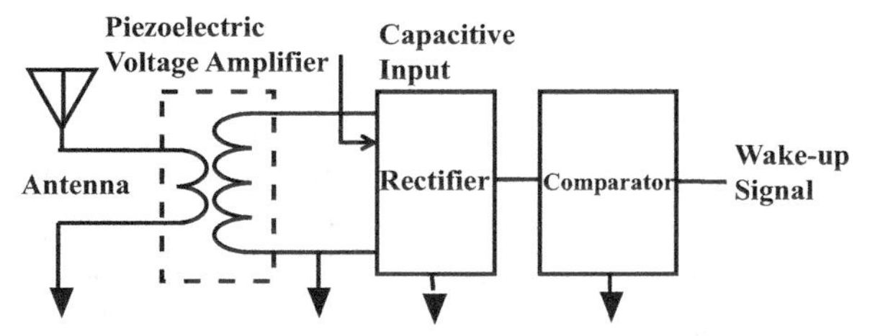

Figure 5.10 Aluminum nitride (AlN) piezoelectric RF resonant voltage amplifier.

Source: Ref. [116].

ratio of the total input area to the total output area, $A_{in}N_{in}/A_{out}N_{out}$, where A_{in} and A_{out} are the areas of the input and output electrodes, respectively, and N_{in} and N_{out} are the number of input and output electrodes/fingers, respectively; these input and output electrodes are being implemented with interdigitated transducers (IDT). The amplification mechanism is discussed next with reference to Lu et al.'s implementation (Figure 5.11) [116].

In the first place (Figure 5.11a), IDTs are patterned on both sides of an AlN substrate. In the second place, an applied voltage, V_{in}, generates a lateral mode acoustic wave via the piezoelectric effect. And, in the third place, the acoustic wave propagates and reflects at the acoustic boundaries, and so, it bounces back and forth in the substrate. The acoustic wave arriving between V_{out} and V_{gnd}, as a result of the input signal between V_{in} and V_{gd}, is characterized by a voltage gain G.

In Figure 5.10, since the rectifier has a capacitive input, the preceding piezoelectric voltage amplifier must produce a high voltage gain across it; this is the situation in most IoT applications, and thus, the open circuit gain is a key figure of merit. Since the system is passive and assumed nearly lossless,

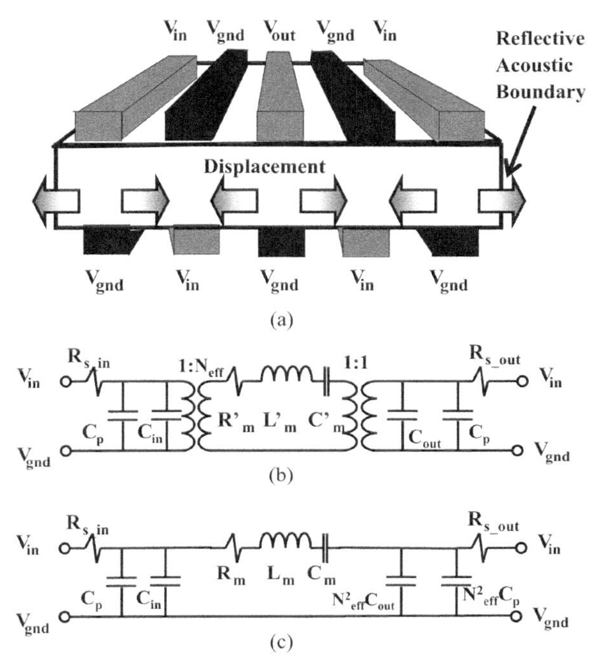

Figure 5.11 AlN resonant voltage amplifier. (a) 3D sketch of the cross-sectional displacement mode shape obtained from FEM. Equivalent circuit models (b) with transformers and (c) without transformers.

Source: Ref. [116].

the input and output powers are equal, that is,

$$I_{in}V_{in} \approx I_{out}V_{out} \tag{5.2}$$

or

$$\frac{V_{out}}{V_{in}} = \frac{I_{in}}{I_{out}} \cong \frac{A_{in}N_{in}}{A_{out}N_{out}} \tag{5.3}$$

The open circuit gain of the device is defined as the ratio of the total voltage on the output port to the total voltage on the input port. This voltage gain may be calculated from the equivalent circuit of the device in Figure 5.11b and is given by [116],

$$G|_{\substack{R_{s\_in}\neq 0\\ \omega=\omega_G}} = \frac{G|_{\substack{R_{s\_in}=0\\ \omega=\omega_G}} \cdot \frac{R_{in}}{R_{s\_in}}}{\sqrt{\left(\frac{R_{in}}{R_{s\_in}}+1\right)^2 + \left(\frac{\pi^2}{8Qk_{t\_in}^2}\right)^2 + \frac{1}{k_{t\_in}^2 Q^2}\frac{\pi^2 N_c}{8N_{eff}^2}}} \tag{5.4}$$

where, in Figure 5.11b, N_{eff} is the effective turns ratio that represents any deviation from the ideal ratio N caused by the distortions in the periodic displacement/stress across the resonator body, C_{in} and C_{out} are the static capacitances of the input and output IDT transducers, and C_p is the total capacitance loading the ports due to the parasitics of the routing lines and the pads. Taking into account these parasitics, the capacitance ratio of the structure is defined as $N_C = (C_{in} + C_p)/(C_{out} + C_p)$, whereas it is assumed that $N_{eff} = C_{in}/C_{out}$. The routing resistances are accounted for via the resistors R_{s_in} and R_{s_out}. After the input circuit, we find a series RLC resonator that represents the piezoelectric resonator (acoustic waves bouncing back and forth laterally), where $\mathrm{R_m}$, $\mathrm{L_m}$, and $\mathrm{C_m}$, are the motional resistance, inductance, and capacitance [81]. These parameters, in turn, are expressed in terms of more fundamental parameters such as the input electromechanical coupling, $k^2_{t\_in}$ [81]; the mechanical resonant frequency, ω_s; and the quality factor, Q, to give,

$$C_m = \frac{8}{\pi^2} C_{in} k^2_{t\_in} \tag{5.5}$$

$$R_m = \frac{1}{\omega_s Q C_m} \tag{5.6}$$

$$L_m = \frac{1}{\omega_s^2 C_m} \tag{5.7}$$

Now, to capture the fact that the device is only partly excited by input transducers, the electromechanical coupling coefficients of the input port, $k^2_{t\_in}$, and output port $k^2_{t\_out}$ are expressed as,

$$k^2_{t\_in} = k_t^2 \cdot \frac{N_{eff}}{N_{eff} + 1} \tag{5.8}$$

$$k^2_{t\_out} = k_t^2 \cdot \frac{1}{N_{eff} + 1} \tag{5.9}$$

where k_t^2 in this case is the intrinsic electromechanical coupling coefficient of an identical device; the excitation of all transducers is derived from the input. Upon analyzing the system, Lu et al. [116] concluded that the maximum gain occurs at the frequency of the series RLC resonance, namely when L_m tunes out C_m and $N_{eff}2C_{out}$. Under these circumstances, for $Q \gg 1$, this frequency

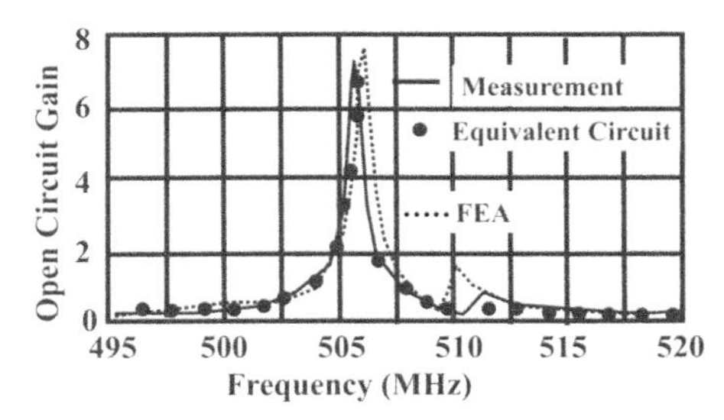

Figure 5.12 Open circuit gain versus frequency. The open circuit gain is defined as the ratio of the total voltage on the output port over the total voltage on the input port.

Source: Ref. [116].

Table 5.5 Voltage amplifier characteristics and equivalent circuit model

Key Characteristics		Equivalent Circuit Model		Performance	
Q^*	1647	N_{eff}	2	Calc. G	6.85
$k^2_{t\_in}$	1.24%	R_m	104 Ω	Meas. G	7.27
$k^2_{t\_out}$	0.62%	L_m	54 μH	Calc. fractional 3 dB BW	0.109%
C_{in}	184 fF	C_m	1.84 fF	Meas. fractional 3 dB BW	0.113%
C_{out}	92 fF	R_{s_in}	82.8 Ω		
C_p	31 fF	R_{s_out}	119 Ω		

Source: Ref. [116].

is approximated by,

$$\omega_G = \omega_s \sqrt{1 + \frac{8}{\pi^2} k_t^2 \frac{1}{N_{eff} + 1} \times \frac{C_{out}}{C_{out} + C_p}} \tag{5.10}$$

Figure 5.12 shows a plot of the open circuit gain, where a peak gain of close to 8 is obtained. This gain, for an input power of −20 dBm, produces a peak output voltage of ~32 mV. The device parameters utilized in the experiment are presented in Table 5.4.

5.3 Mechanical Energy Harvesting

A discussion on mechanical energy harvesting is facilitated by classifying the various sources of mechanical energy. According to Gilbert and Balouchi [118], these sources may be classified into those dependent on motion that remains constant over long periods of time, typified by the air flow used in a turbine; those that derive from an intermittent motion, typified by walking;

and those that derive from cyclic motion, typified by vibration sources. Because of its ubiquity, we will focus on the latter.

A vibration source is characterized by its amplitude and spectrum, that is, the set of vibration frequencies accompanying it. Figure 5.13 shows the acceleration, and Figure 5.14 shows the displacement for vibrations typical of a domestic freezer.

It must be kept in mind that, in general, the amplitude of the vibration and its frequency spectrum are impacted by the presence of the energy harvesting device it feeds, in terms of the relative masses of the harvesting device and that of the vibrating mass [118]. In addition, the available energy may be

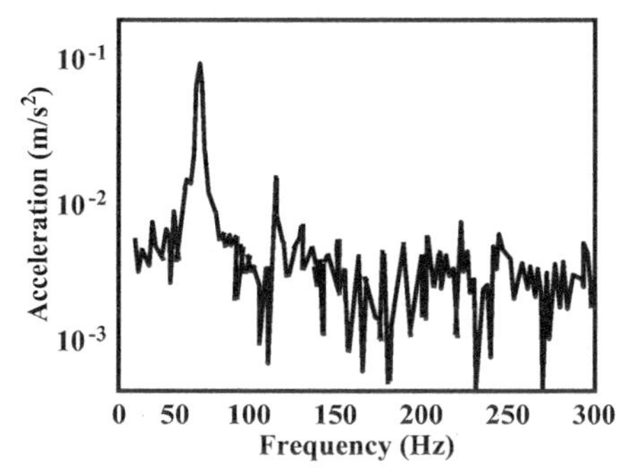

Figure 5.13 Vibration amplitude as a function of frequency for a domestic freezer: Acceleration magnitude.

Source: Ref. [118].

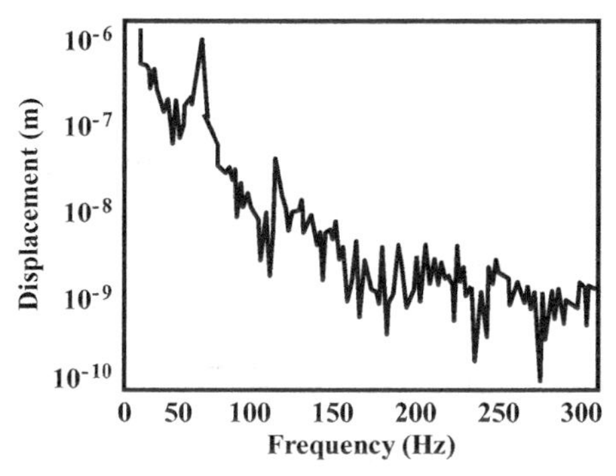

Figure 5.14 Vibration amplitude as a function of frequency for a domestic freezer: Displacement amplitude.

Source: Ref. [118].

altered by changes in the environmental conditions surrounding the vibration equipment, which may affect its vibration modes.

5.3.1 Theory of Energy Harvesting from Vibrations

The theory of vibration-based energy harvesting devices is normally derived by studying the mass, spring, and damper system shown in Figure 5.15. The equation of motion of the system is given by [118],

$$m\frac{d^2x(t)}{dt^2} + b\frac{dx(t)}{dt} + kx(t) = -m\frac{d^2u(t)}{dt^2} \tag{5.11}$$

where b represents the viscous damping coefficient for a damping force proportional to the velocity and $u(t)$ is the displacement.

The steady-state solution to Equation (5.11) is given by [118],

$$x(t) = \frac{\omega^2}{\sqrt{\left(\frac{k}{m} - \omega^2\right)^2 + \left(\frac{b\omega}{m}\right)^2}} U\sin(\omega t + \phi) \tag{5.12}$$

The questions to be asked in determining the performance of the system are the following. For a given displacement of amplitude U and frequency ω, $u(t) = U\sin(\omega t)$: (i) What is the power dissipated? and (ii) What is the useful power extracted/harvested? The answers to these questions have been given by Gilbert and Balouchi [118]. The answer to the first question, the

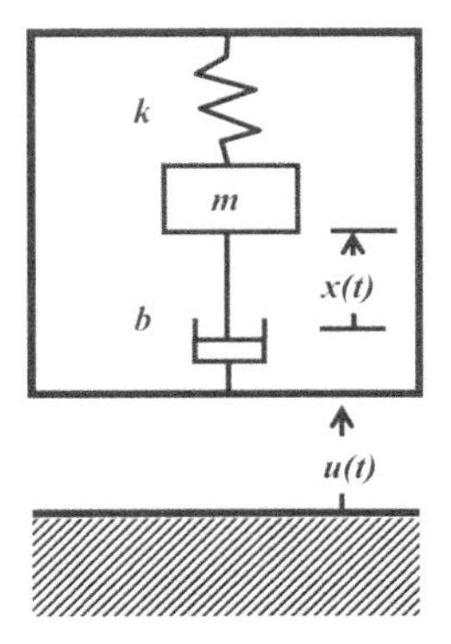

Figure 5.15 Model of a translational inertial generator. The mass is represented by the proof mass m, the spring stiffness by k, and the energy damping by b.

Source: Ref. [118].

power dissipated, is,

$$P_d = \frac{m\xi_T A^2 \omega^2 \frac{\omega^2}{\omega_n^3}}{\sqrt{\left[1-\left(\frac{\omega}{\omega_n}\right)^2\right]^2 + \left[2\xi_T\left(\frac{\omega}{\omega_n}\right)\right]^2}} \tag{5.13}$$

where $\omega = \sqrt{k/m}$ is the mechanical resonance frequency of the system, $\xi_T = b/(2m\omega_n)$ is the total damping factor, and A is the acceleration amplitude. A plot of the normalized power dissipation is shown in Figure 5.16.

As may be appreciated from Figure 5.13, increasing the damping factor causes the peak amplitude, which occurs at a driving frequency equal to the resonance frequency, to decrease and the bandwidth to increase. Thus, in applications in which the system is driven at a constant frequency, a low damping factor is desirable, but in those in which the system is driven by a range of frequencies, higher damping factor would result in higher average power [118].

At resonance, the power dissipated during vibration due to damping is given by,

$$P_d = \frac{mA^2}{4\omega_n\xi_T} \tag{5.14}$$

The second question is obtained by assuming the total damping to be divided into two parts, namely one that represents energy loss due to conversion

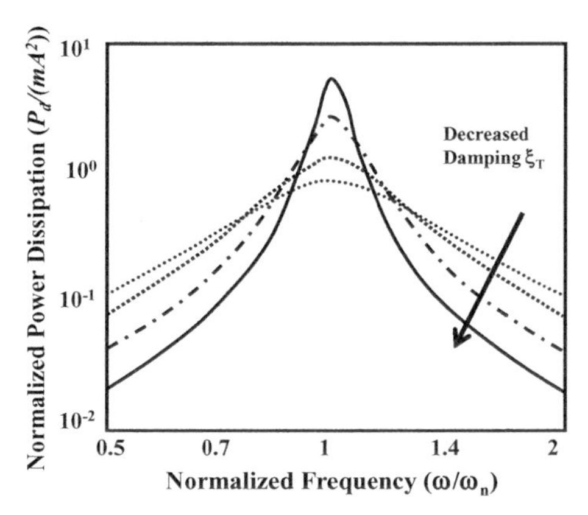

Figure 5.16 Relationship between total dissipated power and frequency for vibration resonator with damping factor as a parameter.

Source: Ref. [118].

from mechanical to electrical energy, ξ_e, and one that represents the parasitic energy losses, ξ_p, representing losses. In terms of these, the useful electrical power is given by [118],

$$P_e = \frac{mA^2\xi_e}{4\omega_n(\xi_e + \xi_p)^2} \tag{5.15}$$

which gives the maximum extracted power when $\xi_e = \xi_p$, namely,

$$P_e = \frac{mA^2}{16\omega_n\xi_e} \tag{5.16}$$

This expression is valid for $\xi_e \neq 0$. It can be shown, however, that as $\xi_e \to 0$, the vibration amplitude would tend to increase indefinitely; however, this increase is limited by practical factors such as the size of the proof mass. When this limitation is taken into account, it was found that the maximum useful power extracted is given by [119],

$$P_{e\,\max} = m^{\frac{4}{3}}\rho^{-\frac{1}{3}}A\omega_n \tag{5.17}$$

where ρ is the density of the proof mass. Equation (5.17) also assumes that the mechanical to electrical transduction mechanism is lossless. However, in reality, there is an energy cost in converting from one form of energy to another. This is captured by the efficiency of the transduction process and, thus, depends of the different approaches to effecting this process. Mechanical vibration to electricity conversion processes are classified as the electromagnetic type, the piezoelectric type, and the electrostatic type. For compact IoT nodes, it is desirable to employ small size energy harvesters, for example, less than 10 cm^3 in volume [123]. Therefore, piezoelectric and electrostatic converters are preferred.

5.3.1.1 Piezoelectric conversion

In piezoelectric mechanical to electrical energy conversion, environmental mechanical vibrations are converted into energy through piezoelectric effect. In particular, with respect to Figure 5.14, as the proof mass vibrates in response to a mechanical acceleration in the z-direction, the expansions and contractions of the piezoelectric layer cause the voltage V to appear (see Section 3.1.2). The proof mass' position at the beam tip results in a larger vibration amplitude and a lower resonance frequency.

A set of analytical expressions for the design of the piezoelectric cantilever resonator, useful for preliminary design, have been given by Saadon

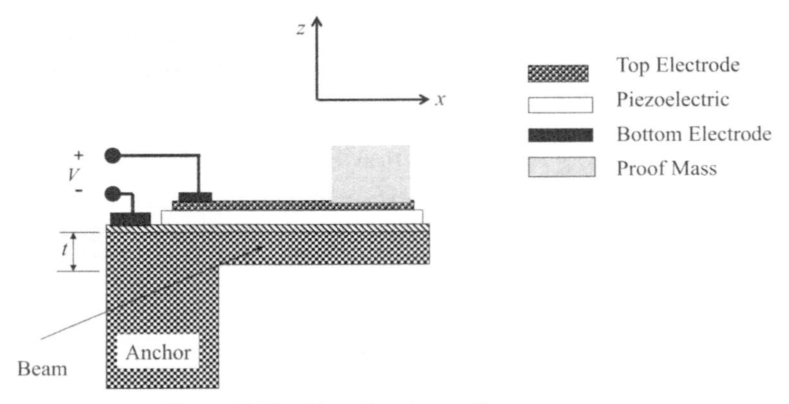

Figure 5.17 Piezoelectric cantilever resonator.

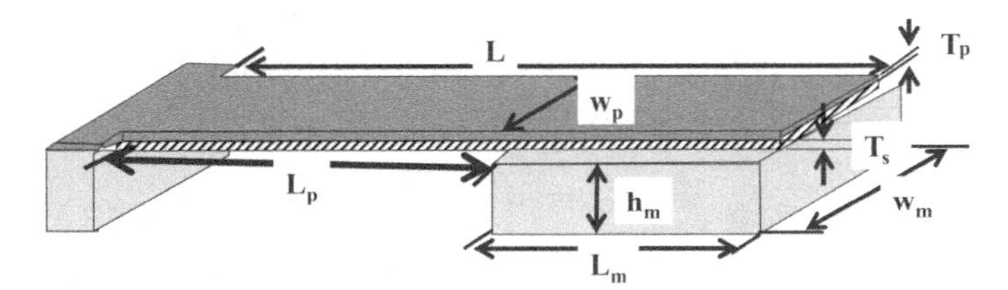

Figure 5.18 Piezoelectric cantilever resonator sketch referred to in design equations.
Source: Ref. [120].

and Sidek [120] and are reproduced below, with reference to the notation in Figure 5.15, for convenience.

Following Saadon and Sidek [120], we proceed as follows. The resonant frequency of a piezoelectric cantilever with a uniform lineal mass density is normally expressed by,

$$f_n = \frac{\nu_n^2}{2\pi l^2}\sqrt{\frac{EI}{m'}} \tag{5.18}$$

where, for the nth mode of vibration, f_n and ν_n^2 are the corresponding resonant frequency and eigenvalues, respectively; l denotes the cantilever beam length; E denotes the cantilever Young's modulus; I denotes the area moment of inertia about the neutral axis; and m' denotes the mass per unit length of the cantilever. Now, due to the composite nature of the beam, it is convenient to write Equation (5.18) in terms of the bending modulus per unit width, D_p. This yields,

$$f_n = \frac{\nu_n^2}{2\pi l^2}\sqrt{\frac{D_p}{m}} \tag{5.19}$$

where $m = \rho_p t_p + \rho_s t_s$. In this way, the mass per unit area, m, is expressed in terms of the sum of the products of the density and thickness of each layer, where $\rho_p t_p$ is the product of the density and thickness of the piezoelectric layer, whereas $\rho_s t_s$ is the product of the density and thickness of the support layer.

For two layers, the bending modulus D_p is a function of the Young's modulus and the thickness of the two layers, and is expressed as,

$$D_p = \frac{E_p^2 t_p^4 + E_s^2 t_s^4 + 2E_p E_s t_p t_s (2t_p^2 + 2t_s^2 + 3t_p t_s)}{12(E_p t_p + E_s t_s)} \tag{5.20}$$

In the case under consideration, the cantilever beam is loaded with a proof mass located at its tip. Therefore, the resonance frequency to take this into account is expressed as,

$$f_r = \frac{\omega}{2\pi} = \frac{1}{2\pi}\sqrt{\frac{k}{m_e}} \tag{5.21}$$

where ω, k, and m_e denote the angular frequency, the spring constant at the tip, and the effective mass of the cantilever, respectively.

Now, when the size of the attached proof mass is smaller than the beam length, the resonant frequency is approximated as,

$$f_n = \frac{\nu_n'^2}{2\pi}\sqrt{\frac{k}{m_e + \Delta m}} \tag{5.22}$$

where the effective mass $m_e = 0.236\ mwl$ when considering the axial velocity, which acts on the length or the width $(w \ll l)$. In this case the spring constant, k, for the composite beam may be expressed as,

$$k = \frac{3D_p w_p}{l^3} \tag{5.23}$$

On the other hand, whenever the proof mass is not located at the tip of the beam, but its center is concentrated at a distance $l_m/2$ from the tip, then the effective spring constant at this point is expressed as,

$$k' = k\left(\frac{l}{l - \frac{l_m}{2}}\right)^3 \tag{5.24}$$

With these considerations, the resonance frequency for the piezoelectric cantilever energy harvester is given by [120],

$$f_n = \frac{\mathrm{v}_n^2}{2\pi}\sqrt{\frac{0.236 w_p D_p (l - l_m/2)^3}{0.236 m w_p l^7 + \Delta m l^3 (l - l_m/2)^3}} \tag{5.25}$$

The energy conversion process in a piezoelectric cantilever energy harvester is characterized by an efficiency, η. In particular, the conversion process consists in that, when mechanically strained, the composite piezoelectric beam will generate across it an electric field proportional to the mechanical strain, whereas, when experiencing an applied electric field across it, a mechanical strain is generated in it. These phenomena are described by the piezoelectric strain constant, d, which gives the relationship between applied mechanical stresses and the resulting electric field, and the electromechanical coupling coefficient, k, which relates the applied electric field to the resulting mechanical strain. This latter coefficient enters determining the efficiency of a piezoelectric resonant generator [119]. For a clamped–clamped piezo element that is cyclically compressed at its resonant frequency, the efficiency is [119],

$$\eta = \frac{\frac{k^2}{2(1-k^2)}}{\frac{1}{Q} + \frac{k^2}{2(1-k^2)}} \tag{5.26}$$

where Q is the quality factor of the resonator. Obviously, k is a property of the piezoelectric material being utilized. Typical materials and their respective k values are lead zirconate titanate (PZT), $k \leq 0.75$; and polyvinylideneflouride (PVDF), $k \sim 0.12{-}0.15$.

A recent MEMS/NEMS implementation of the piezoelectric cantilever energy harvester was reported recently by Iannacci et al. [121] (Figure 5.19). In this device, proof masses are located at the extremes of a thin beam/membrane that is anchored as shown (Figure 5.19). The masses are linked to the thin central portion of the membrane via stiffening wedges. As the beam vibrates, the normal in-plane strain, required for piezoelectric conversion, is confined to the central area of the membrane, that is, where an AlN piezoelectric conversion layer is located. By connecting electrodes to the anchors, as shown in Figure 5.20, the voltage developed via the piezoelectric effect is sensed.

The analysis and design of the device is best carried out using a finite-element analysis (FEA) software tool that numerically simulates its behavior. Figure 5.21 shows simulated and measured results for the output power versus frequency of the device [121].

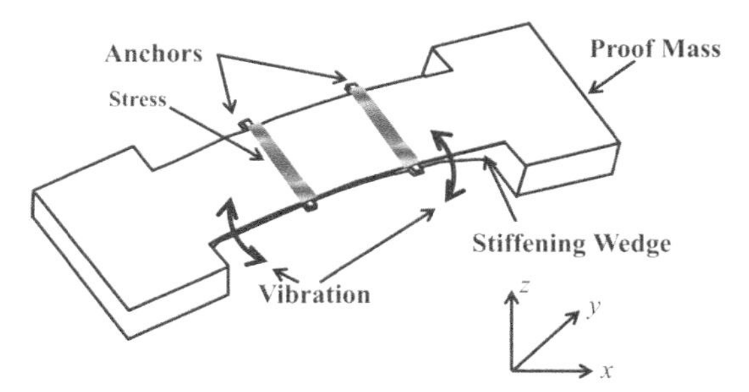

Figure 5.19 Piezoelectric MEMS/NEMS energy harvester: 3D flexing view. A central rectangular membrane, made of 20-μm-thick silicon, is fixed to the surrounding frame with four small anchors.

Source: Ref. [121].

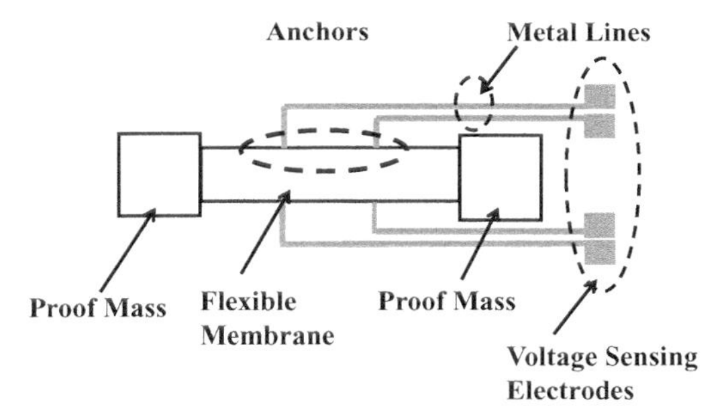

Figure 5.20 Piezoelectric MEMS/NEMS energy harvester: Top view showing interface to electronics.

Source: Ref. [121].

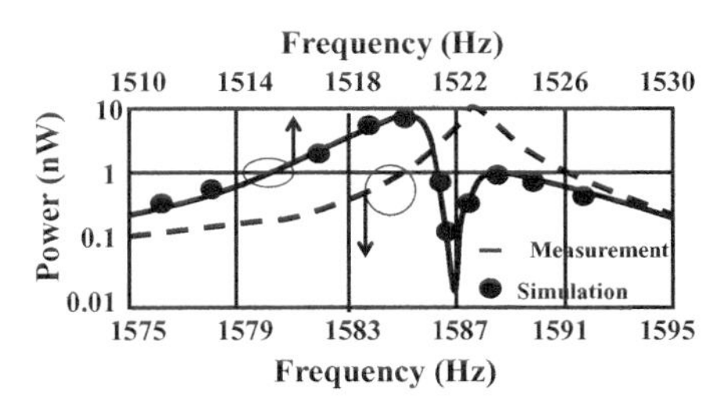

Figure 5.21 Measured versus simulated output power of the piezoelectric energy harvester loaded with a 1 MΩ resistor at resonance and stimulated with an acceleration of 1.5 m/s^2.

Source: Ref. [121].

5.3.1.2 Electrostatic conversion

In the electrostatic conversion paradigm, power is extracted from a *charged* capacitor, while its capacitance varied as a result of mechanical vibration. Since the capacitance is given by,

$$C = \frac{\varepsilon A}{d} = \frac{Q}{V} \tag{5.27}$$

the capacitance may be caused to change when the vibration alters its inter-plate (inter-electrode) dielectric constant, ε, or separation, d, or its area, A, which, in turn, manifests as a change in its charge, Q, its voltage or both [122]. From the fact that, as discussed in Section 3.1.1 for a parallel-plate capacitor, the opposite charges in the plates produce a force of attraction that tends to reduce d, and when the plates are constrained to be fixed/unmovable, energy is stored of a value,

$$E = \frac{1}{2}QV \tag{5.28}$$

which may be expressed as [118],

$$E = \frac{Q^2V}{2\varepsilon A} \tag{5.29}$$

when the charge is constrained to be constant, or,

$$E = \frac{\varepsilon AV^2}{2d} \tag{5.30}$$

when the voltage is constrained to be constant. The electrostatic conversion processes may be understood with reference to Figure 5.22, as explained by Boisseau et al. [123].

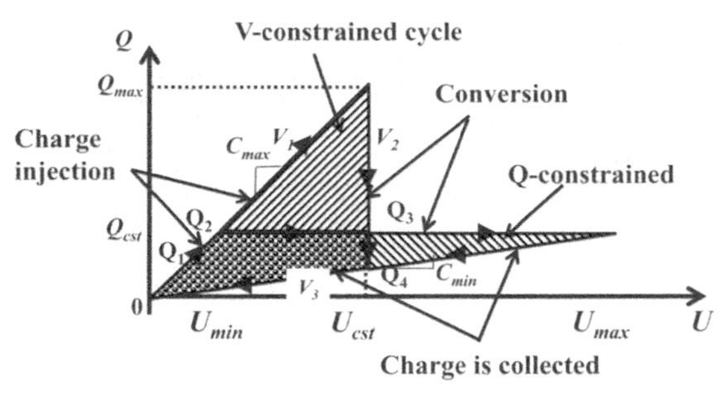

Figure 5.22 Standard energy conversion cycles for electret-free electrostatic devices.

Source: Ref. [123].

In general, there are two approaches to effect electrostatic conversion, depending on whether or not electrets are employed. An electret is "a dielectric material that has a quasi-permanent electric charge or dipole polarization that generates internal and external electric fields, and is the electrostatic equivalent of a permanent magnet" [124]. In the electrostatic converter approach that does not use an electret (the electret-free approach), use is made of an active electronic circuit that requires an external agent, for example, a battery, a charged storage capacitor or the like, to effect charging/discharging cycles of the capacitance and must be synchronized with the capacitance variation. In the context of realizing fully autonomous IoT nodes, the need for a battery would render autonomy impossible. On the other hand, in the approach that uses the electret, it is possible to directly convert mechanical power into electrical power without the need for an external battery.

The electret-free approach to electrostatic conversion may be effected in either of the two cycles, namely the charge-constrained cycle and the voltage-constrained cycle.

The charge-constrained cycle (Figures 5.22 and 5.23a) starts when the charge Q_1 is injected into the device until it reaches its maximum capacitance C_{max}. During this path, the capacitor is being charged by an external source until an electric charge Q_{cst}, corresponding to a voltage U_{min}, is stored. The device is then placed into the open circuit state Q_2. Next, the device moves mechanically to position Q_3, where its capacitance is at a minimum. Since the charge Q_{cst} is held constant while the capacitance C decreases, the

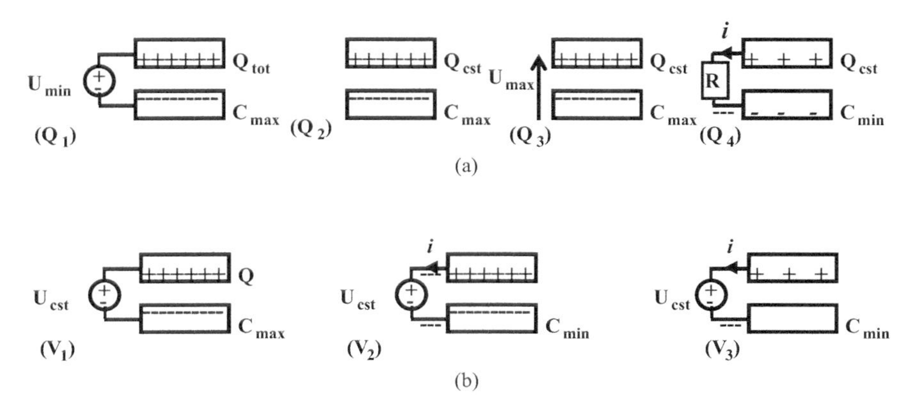

Figure 5.23 Standard energy conversion cycles for electret-free electrostatic devices. (a) Charge-constrained cycle. (b) Voltage-constrained cycle.

Source: Ref. [123].

voltage across the capacitor U increases. Then, when the capacitance reaches its minimum, C_{min}, or the voltage its maximum, U_{max}, electric charges are removed from the structure at state Q_4. The total amount of energy converted during each cycle is given by [123],

$$E_{Q=cte} = \frac{1}{2}Q_{\text{cst}}^2 \left(\frac{1}{C_{\min}} - \frac{1}{C_{\max}} \right) \tag{5.31}$$

The voltage-constrained cycle (Figures 5.22 and 5.23b) starts when the capacitance of the device is at maximum at V_1. Following path V_1, the capacitor is charged by an external voltage source until a voltage U_{cst} is attained. This voltage is maintained throughout the conversion cycle by an external electronic circuit. Since the voltage is constant and the capacitance decreases, during the vertical V_2 path, the charge of the capacitor decreases, thus generating a current that is extracted from the capacitor. When the capacitance reaches its minimum value, the charge Q still present in the capacitor is completely collected during path V_3. The total amount of energy converted during each cycle is given by,

$$E_{U=cte} = U_{cst}^2(C_{\max} - C_{\min}) \tag{5.32}$$

Boisseau et al. [123] have pointed out that, to attain a high efficiency, it is necessary to employ a voltage of the order of 100 V or greater. This, therefore, sets a limitation on these devices, namely that they require an external supply in order to initiate the first cycle. After the first cycle, they claim that part of the energy harvested at the end of a cycle can be re-injected into the capacitor to start the next cycle. The need for an external battery may be circumvented by the utilization of electrets [124]. These, electrically charged dielectrics enable the permanent polarization of electrostatic energy harvesters, thus a direct mechanical-to-electrical conversion.

We conclude this section with an interesting electrostatic energy harvester concept advanced by Huang et al. [125] (Figure 5.24). In this concept, the externally applied voltage of the electret-free electrostatic converter is replaced by the charge generated by a coating of photosensitive material on an RF MEMS capacitor (Figure 5.25). In operation, the charge developed for a switch of capacitance C is related to the developed voltage V across it by $Q = C \times V$. When the charge accumulated in the capacitance is such that the pull-in voltage is reached, the beam snaps down and the charge flows out of the capacitor into a storage component where a voltage of 2.7–4.2 V is stored. This is, in turn, applied to a regulator that drives the load and the charging cycle repeats.

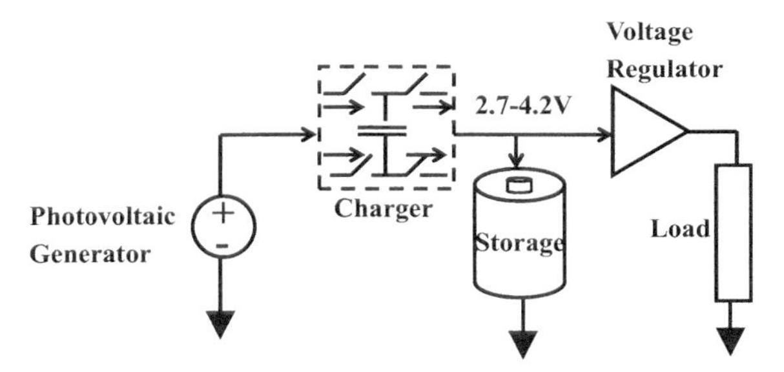

Figure 5.24 Representation of RF MEMS energy harvester.

Source: Ref. [125].

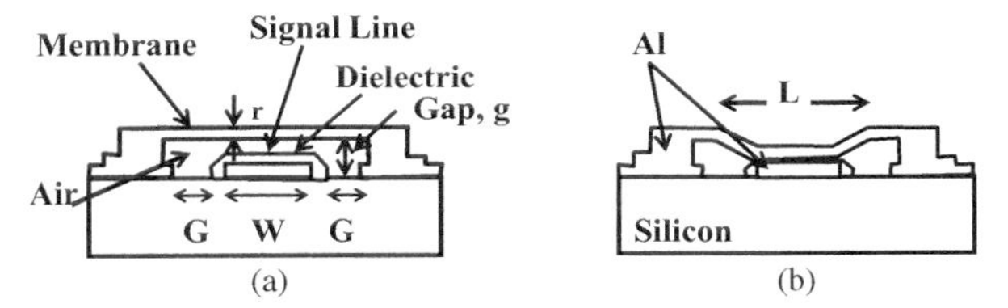

Figure 5.25 Sketch showing the operation of an RF MEMS switch. (a) Beam/membrane UP. (b) Beam/membrane DOWN.

Source: Ref. [125].

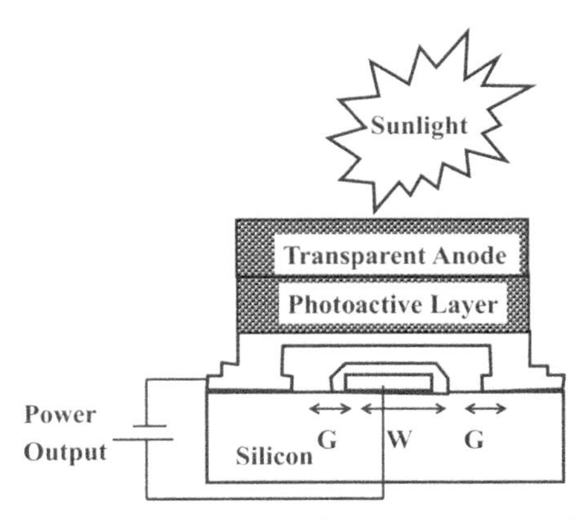

Figure 5.26 Sketch showing the operation of an RF MEMS switch: The structure and operation of the energy harvester. The application of a photosensitive coating and transparent electrode on the beam causes an electric charge to be generated and stored in the RF MEMS structure.

Source: Ref. [125].

Table 5.6 Dimensions of the RF MEMS switch

Component	Material	Dimension
MEMS bridge	Aluminum	t = 1 μm L = 150 μm w = 200 μm
Signal line	Aluminum	Thickness = 1 μm W = 100 μm
Dielectric layer	SiO_2	$g_{SiO_2} = 1$ μm
Air gap	Air	$g_{\mathrm{Air}} = 1.7$ μm
GSG gap	Air	$G = 25$ μm

Source: Ref. [125].

The concept was experimentally implemented using the capacitance of an RF MEMS switch (Figure 5.23), which uses aluminum as the beam's structural material, with a Young's modulus $E = 70$ GPa and a spring constant $k = 132.7$ N/m, resulting in a pull-in voltage of $V_p = 40$ V; its dimensions are described in Table 5.6.

While the concept was not fully demonstrated experimentally, it shows great potential for powering autonomous IoT nodes.

5.4 Summary

In this chapter, we have dealt with the concept of "Energy Harvesting." In particular, the fundamentals of various approaches to power autonomous IoT nodes have been presented. The approaches discussed are: (i) Wireless energy harvesting, dealing with the capture and conversion into DC power of RF signals present in the environment, including radio and TV signals, and cell phone base station signals; (ii) Mechanical energy harvesting, dealing with the piezoelectric and electrostatic approaches. After discussing the fundamentals, examples of implementation of these approaches were also presented.

6

NEMX Applications in the IoT Era

6.1 Introduction

Up to this point, we have dealt with the understanding of nanoelectromechanical quantum circuits and systems (NEMX), as exemplified by: (1) Uncovering of their origins, impetus, and motivation; (2) Developing an understanding of their device physics, including, the topics of actuation, mechanical vibration, and sensing; (3) Studying the fundamentals of key devices, namely, MEMS/NEMS switches, varactors, and resonators, including implementations gathered from the literature; (4) And studying their energy supply via energy harvesting, in particular, as derived from wireless energy and mechanical vibrations. In this chapter, after an introduction of the fundamentals of IoT networks and nodes, we explore how the NEMX components are encroaching in a variety of IoT applications. Finally, we address the subject of the IoT in the context of emerging mmWaves/5G (fifth generation wireless networks) technology.

6.1.1 Wireless Connectivity

The fundamental physical phenomenon that enables the IoT is the propagation of electromagnetic waves between network *nodes*, which occurs at certain frequencies. These frequencies may be conceived of as pertaining to licensed bands or unlicensed bands. License bands are frequencies whose utilization requires authorization for their use, that is, a license from governments of respective countries. On the other hand, unlicensed bands are frequencies whose use is unregulated, thus, everyone is free to use them without government authorization, that is, without a license, provided certain limits regarding acceptable transmitted power and interference levels are complied with. While the prototypical example of licensed bands are those frequencies employed for cell phone communications, that for unlicensed

Table 6.1 Worldwide unlicensed frequency bands

USA	Europe	Japan	Africa	Asia	Rest of the World
• 315 MHz	• 433 MHz	• 315 MHz	• 422 MHz	• 315 MHz	• 2.4 GHz
• 433 MHz	• 868 MHz	• 426 MHz		• 433 MHz	• 5 GHz
• 915 MHz		• 950 MHz		• 470 MHz	
				• 780 MHz	
				• 433 MHz	
				• 915 MHz	

Source: Ref. [126].

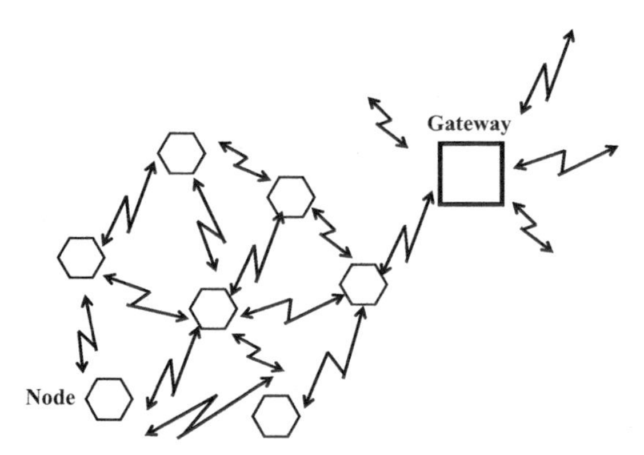

Figure 6.1 Network topologies: Mesh.

Source: Ref. [126].

bands is typified by those used for industrial, scientific, and medical applications, the so-called "ISM" applications. Table 6.1 lists unlicensed frequencies throughout the world [126].

Since attaining the highest possible propagation range is desirable (it reduces the needed transmitter power consumption and allows a larger network), this is one of the key criteria behind the selection of the frequency bands utilized. In this context, there is a trade-off in the sense that, while bands at higher frequencies permit a greater number of channels with greater bandwidth and, thus, greater data rate, the lower frequencies propagate further, as well as achieving wider coverage, in particular, within buildings where non-line-of-sight propagation dominates.

Once the frequencies for the applications are established, the IoT, as a network, must be designed to wirelessly connect to a gateway, or point of connection, to the internet, and to the various *nodes* making up the network. The network is normally configured in either of the two topologies, namely the mesh topology or the start topology (Figures 6.1 and 6.2).

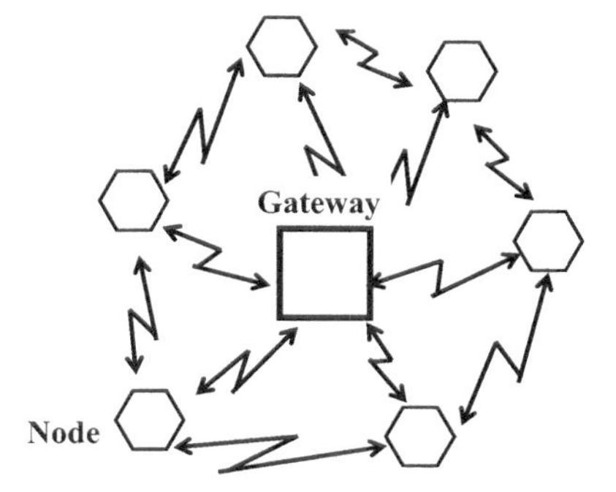

Figure 6.2 Network topologies: Star.

Source: Ref. [126].

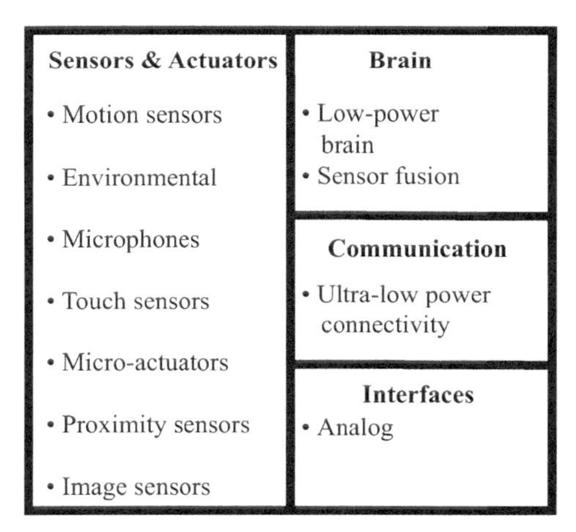

Figure 6.3 STMicroelectronics building blocks for IoT *node*.

Source: Ref. [127].

One can conceive of many realizations for the nodes; Figure 6.3 shows the constituents of a typical IoT node as envisioned by STMicroelectronics [127].

Figure 6.3 reveals that the key enabler to electromagnetic (EM) wave-mediated network communications is the communication building block. The implementation of this building block is dictated by the *standard* utilized to regulate the EM interactions. Table 6.2 shows the key standards currently utilized and their descriptions.

The standards, which are rules that set the formal procedures for networks to operate within them and with one another, are discussed next.

Table 6.2 Some of the key considerations that will influence the choice of wireless protocols for a specific application, such as data rate, range, and power

	Wi-Fi*	BLE/ Bluetooth 5	Thread	Sub-1 GHz: TI 15.4	Sub-1 GHz: Sigfox	Zigbee
Data throughput	Up to 72 Mbps	Up to 2 Mbps	Up to 250 kbps	Up to 250 kbps	100 bps	Up to 250 kbps
Range**	100 m	Up to 750 m	100 m via mesh	4 km	25 km	130 m LOS
Power consumption	Up to 1 year on AA batteries	Up to years on a coin-cell battery	Up to years on a coin-cell battery	Up to years on a coin-cell battery for 1 km range	Up to years on a coin-cell battery for limited range	Years on a coin-cell battery
Topology	Star	Point-to-point/mesh	Mesh and star	Star	Star	Mesh and star
IP at the device node	Yes	No	Yes	No	No	No
PC, mobile OS support	Yes	Yes	No	No	No	No
Infrastructure widely deployed	Yes, Access points	Yes, smart phones	No	No	No	No

*Single stream 802.11n Wi-Fi MCUs may support lower throughput than peak physical capacity of the network.
**LOS = line of sight. For range, note that maximum data rates are often not available at the longest range.
Source: Ref. [126].

Application Layer	HTTP
Transport Layer	TCP
Network Layer	IP
Link Layer	Wi-Fi®

Figure 6.4 Simplified OSI model (left) and an example of a TCP/IP protocol stack (right). The TCP/IP stack is a complete set of networking protocols.

Source: Ref. [126].

6.1.1.1 Communication protocols

The groundbreaking opportunity enabled by the IoT paradigm is that, in principle, local networks may be endowed with a worldwide reach through their links to the Internet. From this point of view, local network nodes may be modeled by the Open Systems Interconnection (OSI) model (Figure 6.4) [126]. In this model, communication is broken into functional layers that enable the easier implementation of scalable and interoperable (i.e., networks operating with different standards can communicate between them) networks.

The *Technology UK* website [128] has accessible definitions for a number of terms displayed in Figure 6.4. The "Transmission Control Protocol (TCP)" and the "Internet Protocol" (IP), or "TCP/IP" is defined as "the world's most widely-used non-proprietary protocol suite because it enables computers using diverse hardware and software platforms, on different types of networks, to communicate." In particular, the TCP/IP is a collection of protocols that facilitate common applications such as *electronic mail*, *terminal emulation*, and *file transfer* [128].

Other components of the OSI model in Figure 6.4 are as follows [128]:

(1) The Link Layer—is usually broken down into two sub-layers for wireless protocols, namely, the logical link control (LLC) layer and the media access control (MAC) layer. The Link layer is responsible for converting bits to radio signals, and vice versa, framing the data for reliable wireless communication, and managing access to the radio channel. In the TCP/IP collection of protocols of Figure 6.3, Wi-Fi is the Link layer protocol;

(2) The Network Layer—Addresses and routes data through the network, with IP being the internet's network layer protocol, providing an IP address to devices and carrying IP packets from one device to another;

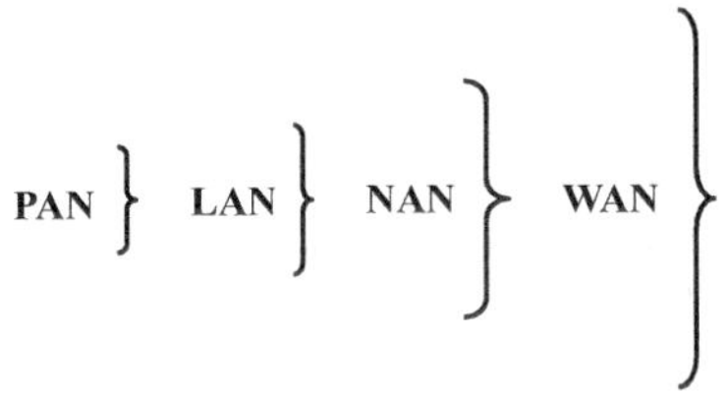

PAN LAN NAN WAN

Figure 6.5 Various ranges and applications for PANs, LANs, NANs and WANs.
Source: Ref. [126].

(3) The Transport Layer—generates communication sessions between applications running on two ends of a network. It allows multiple applications to run on one device, each using its own communication channel. TCP is the internet's predominant transport protocol;
(4) The Application Layer—is responsible for data formatting and governs the data flow in an optimal scheme for specific applications. A popular application layer protocol in the TCP/IP stack is Hypertext Transfer Protocol (HTTP), which was created to transfer web content over the internet.

6.1.1.2 Network range

As mentioned previously, the IoT enables a local network's range to be extended from the smallest range, namely, the personal area network (PAN), connecting instruments monitoring the human body, to local area networks (LANs), connecting in-home appliances, to neighborhood area networks (NANs), connecting houses and buildings in a city, to wide area networks (WANs), connecting countries throughout the world (Figure 6.5).

The various network ranges may be quantified as shown in Table 6.3.

6.2 Roots of the Internet of Things

It could be said that one of the first pioneering works of relevance to today's IoT was the wireless communication system for sending and receiving data from distributed sensor networks in the context of the Smart Dust network concept, introduced by Pister et al. [129] circa 1998. The Smart Dust concept was defined as a set of millimeter scale sensing and communication platforms. The set of platforms working together (communicating amongst themselves via wireless links) numbered hundreds to thousands of dust "motes" in conjunction with one or more interrogating transceivers. Each

Table 6.3 Network ranges, applications, and standards

Network	Range	Typical Applications	Wireless Standard
PAN	10 m	• Wireless headset • Watch or fitness device	• Bluetooth®
LAN	100 m	• Personal computers • Smart phones • TVs • Thermostats • Home appliances	• Wi-Fi
NAN	25 km	• Smart grid network that transmits electric meter readings from homes to utility companies	• Proprietary protocol over a 900 MHz radio
WAN	Spread across a very large area – as big as the entire globe	• The Internet is a complex mix of wired and wireless connections	• Encompasses all standards

Source: Ref. [126].

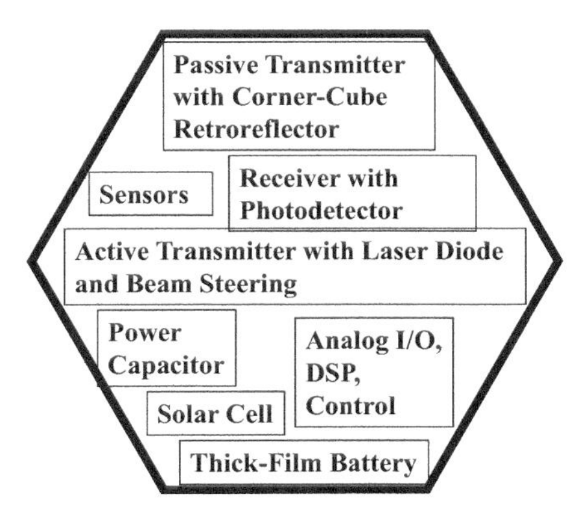

Figure 6.6 Elements of Smart Dust mote.

dust could be considered as a node in a distributed sensor network system, consisting of a power supply, a sensor or sensors, analog and digital circuitry, and a system for receiving and transmitting data (Figure 6.6)

The nodes in the IoT may be conceptualized as a generalization of the motes in Smart Dust, to extend to a small local network to a virtually unlimited number of nodes encompassing up to trillions of nodes and spreading over the globe via internet links. Next, we survey emerging IoT applications and the opportunities for MEMS/NEMS in them.

6.3 Applications of the Internet of Things

The applications of the IoT are virtually unlimited. It is useful in studying them, however, to organize them in terms of the following realms [130]: (1) Smart Home; (2) Wearables; (3) Smart City; (4) Smart Grid; (5) Industrial Internet; (6) Connected Car; (7) Connected Health; (8) Smart Retail; (9) Smart Supply Chain; (10) Smart Farming. The popularity of these applications is ranked in Figure 6.7 [130].

Next, we place MEMS/NEMS in the context of their opportunities for insertion into these applications.

6.3.1 NEMX in Smart Home IoT Applications

The goal of devices for the Smart Home is to facilitate the automation of the home, providing homeowners with peace of mind, convenience and energy efficiency by allowing them to control smart devices, often by a smart home

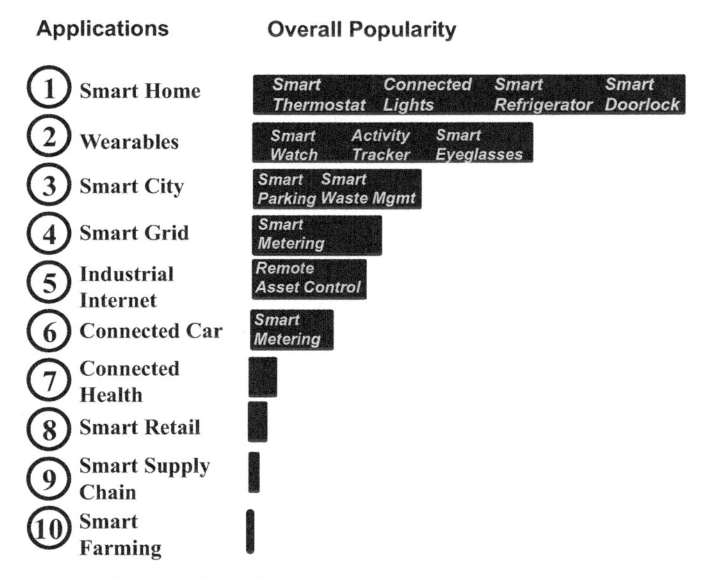

Figure 6.7 Relative popularity of IoT applications.

Source: Ref. [130].

Table 6.4 Opportunities for MEMS/NEMS in the Smart Home

Smart Home IoT Applications	Devices					
	Switch	Varactor	Resonator	Accelerometer	Vibration Sensor	Energy Harvester
Smart Door Locks	X	X	X			
	In this application, users have the ability to control smart locks and garage door openers remotely, to either grant or preclude access to visitors. In addition, smart locks can spontaneously unlock doors upon perceiving the presence of the home owners nearby.					
Smart Thermostats	X	X	X			
	In this application, users are allowed to program the temperature profile in their homes to program, monitor and remotely set up throughout desired temperature profiles. In addition, these smart devices can infer homeowners' preferences and establish desired settings to provide utmost comfort while minimizing cost.					
Smart Security Cameras	X	X	X			
	In this application, people can maintain their residences under surveillance while away, for example, on travel. Together with smart motion sensors, security may be enabled to differentiate among various types of visitors, for example, pets and burglars, and based on this send notification to the police, if appropriate.					
Smart TV	X	X	X			
	Here, TVs are enabled to connect to the internet and search for a variety of programming such as video on-demand and music. In addition, it may be possible for smart TVs to also possess the ability to recognize voice and/or gestures.					
Smart Watering	X	X	X			
	In this application, the watering of plants and lawns can be programmed to schedule the opening and closing of water valves automatically at desired times and durations.					
Smart Kitchens	X	X	X			
	In this application, a wide variety of appliances such as smart refrigerators, slow cookers, and coffee makers can be programmed to alert regarding the proximity of food expiration dates, In addition, smart washing and drying machines could be made that sense the types of clothes put in them and effect pertinent washing and drying cycles accordingly.					

Source: Ref. [131].

application ("app") on their smart phone or other networked device. As part of the IoT, the devices and systems in a smart home usually operate in concert, sharing data among themselves and enabling the automation of actions based on the homeowners' programming.

Table 6.5 Opportunities for MEMS/NEMS in wearables

Wearables IoT Applications	Devices					
	Switch	Varactor	Resonator	Accelerometer	Vibration Sensor	Energy Harvester
	X	X	X			
ECG and tilt	Measurement of blood pressure and heart rate while exercising, for example, running.					
			X	X	X	
SpO_2 and motion sensors	Measurement of blood oxygen saturation (SpO_2) while exercising. A pulse oximeter also measures the pulse rate simultaneously with SpO_2.					
			X			
Temperature	Monitoring body temperature and replaying it to a central station for storage and further analysis.					
			X	X	X	
Pulse	Measurement of pulse rate while exercising and relaying it to a central station for storage and further analysis.					

Source: Ref. [132].

6.3.2 NEMX in Wearable IoT Applications

Wearable devices usually consist of the following elements: (1) A layer of sensors that are placed in contact with the body to monitor parameters such as temperature, movement and pulse. In addition, they include a connectivity layer, usually employing the Bluetooth Low Energy (BLE) protocol to connect to a smart phone or home network. Finally, it also interacts with the cloud with which the wearable exchanges data. Table 6.5 shows opportunities for MEMS/NEMS devices in wearables.

6.3.3 NEMX in Smart Cities IoT Applications

The motivation behind bringing the IoT to bear on cities is rooted on the fact that, according to certain studies [133], by the year 2050 it is expected that close to 70 percent of the world's population will move into cities. This population increase is expected to drastically strain cities' infrastructure. In particular, a number of issues, from traffic management, to waste management, to water metering and distribution, to security and monitoring the environment will become important to enable life in cities. IoT is aimed at impacting the areas of traffic congestion, reduction of pollution and noise, and increasing personal security [133, 134]. Table 6.6 shows opportunities for MEMS/NEMS devices in Smart Cities.

Table 6.6 Opportunities for MEMS/NEMS in Smart Cities

Smart Cities IoT Applications	Devices					
	Switch	Varactor	Resonator	Accelerometer	Vibration Sensor	Energy Harvester
Smart Water Meters	X	X	X			
	This application can enable, not only remote metering and billing, thus reducing cost and enhancing resource management to the utilities company, but also affording utilities company greater visibility on customers' consumption profiles.					
Smart Traffic Signals	X	X	X	X	X	X
	This refers to exploiting various kinds of sensors, including obtaining the GPS data in driver's smart phones, to determine traffic volume and distribution, and vehicle speeds; then, using this information to control traffic lights to speed up traffic and preclude congestion.					
Smart Building Energy Management	X	X	X			
	The goal in exploiting IoT for building energy management is to bring recent energy saving techniques to older buildings. In particular, this will enable interfacing modern approaches to legacy heating, cooling, lighting, and fire-safety systems.					
Smart Video Monitoring Systems	X	X	X	X	X	X
	A smart network of cameras distributed throughout the city enables effecting surveillance to promote citizen's safety and security. The flexibility afforded by autonomous IoT nodes enabling reconfigurable networks would be a new twist that will enhance the flexibility of security plans.					

Source: Ref. [134].

6.3.4 NEMX in Smart Grid IoT Applications

A Smart Grid is defined as: "An electricity network that can intelligently integrate the actions of all users connected to it—generators, consumers and those that do both—in order to efficiently deliver sustainable, economic and secure electricity supplies" [136]. The motivation behind bringing the Smart Grid to the IoT stems from the fact that it is desirable to exploit the information pertaining to energy delivery and consumption patterns of power plant supplying electricity, on the one hand, and consumers, on the other hand, in a real time fashion to automatically control the efficiency, reliability, and cost of electricity.

6.3.5 NEMX in Industrial Internet IoT Applications

The term Industrial Internet is a term originated by General Electric [137]. It is aimed at applications involving the interconnection of machines and devices in industries such as oil and gas, power utilities and healthcare.

Table 6.7 Opportunities for MEMS/NEMS in Smart Grid

Smart Grid Applications	Devices					
	Switch	Varactor	Resonator	Accelerometer	Vibration Sensor	Energy Harvester
Smart Meter	X	X	X			
	Here, the collection of data from nodes distributed throughout home appliances can be exploited to obtain energy usage profiles that, linked to Smart Meters, can be coordinated with the Smart Grid to balance the consumption and availability schedules of both users and utility companies.					
Home Energy Management	X	X	X	X	X	X
	In this area, consumers can utilize the IoT to manage their energy consumption profiles in concert with the real time electricity price, in view of total demand, to achieve cost savings.					
Online Monitoring and Self-Healing	X	X	X		X	
	Here, the IoT network can be configured to have nodes distributed along various power grid aspects such as the power plant, transmission lines, distribution lines, energy consumption, and energy storage. This information can then be exploited to improve the Smart Grid's ability to self-heal. In particular, the utilization of sensors to detect malfunctions in the system will enable its ability to respond accordingly.					

Source: Ref. [136].

The implementation of the Industrial Internet relies on the networking of instrumentation, sensors and devices to machinery and vehicles engaged in the transport, energy and industrial markets, in particular, oil and gas, power utilities and healthcare.

The Industrial IoT integrates a number of emerging paradigms such as machine learning and big data technology, thus embodying a medium harnessing data deriving from sensors a plethora of machine-to-machine (M2M) communication and automation technologies employed in industrial processes. In this fashion, the efficiency of processes can be increased, leading to time and money savings [138].

6.3.6 NEMX in Connected Car IoT Applications

With the increasing proliferation of electronics in cars, including, Internet connectivity, the number of opportunities for connecting cars to both the Internet and other cars, and car-related services, is set to explode. Modern cars come now equipped by either a built-in antenna and chipset for Internet connectivity or a hardware that allows drivers to connect to their cars via their smart phones. Below, we list a number of connected car applications.

Table 6.8 Opportunities for MEMS/NEMS in Industrial Internet

Industrial Internet Applications	Devices					
	Switch	Varactor	Resonator	Accelerometer	Vibration Sensor	Energy Harvester
Smart Factory Warehousing Applications	X	X	X			
	In the context of a warehouse, devices can be deployed to enable both humans and machines to integrate their activities to yield increases in productivity and workflow. In particular, this can be achieved by connecting portable computers, barcode scanners, radio-frequency identification (RFID) readers, wearables, sensors, and aisle and shelf beacons.					
Predictive and Remote Maintenance	X	X	X	X	X	X
	The deployment of machine vibration sensors enabled with wireless connectivity can monitor the health of machines and help avoid catastrophic events.					
Freight, Goods and Transportation Monitoring	X	X	X			
	Here, the clear increase in online shopping, in addition to the traditional movement of freight and goods, justifies the exploitation of the IoT to monitor a fleet of trucks and, enable the future paradigm shift towards autonomous trucks.					
Smart Farming and Livestock Monitoring	X	X	X	X	X	X
	As in other endeavors employing distributed assets, the IoT can facilitate increases in crop productivity and product quality, by helping farmers to better conserve their resources and controlling cost. Towards this end, for instance, real time data gathering pertaining soil quality and moisture levels can play crucial roles in harvest optimization by, for example, providing the tools for elucidating the right parameters for plant growth, predicting pest behavior and managing irrigation requirements. Similarly, livestock wearables can include sensors to monitor a variety of vital signs such as their heart rate, blood pressure, temperature, etc.					
Industrial Security Systems	X	X	X	X	X	X
	The high degree of connectivity enabled by the IoT is ideal for enabling ubiquitous monitoring or surveillance of all aspects of a facility. In particular, by establishing automatic interactions with law enforcement agencies, the large scale nature of industrial security systems may be protected in real time.					
Industrial Heating, Ventilation and Air Conditioning	X	X	X	X	X	X
	In this realm, the IoT brings the ability to embed smart nodes deep into building infrastructure, in particular, the equipment providing heating, ventilation, air conditioning and refrigeration. These smart nodes, connected to the Internet, would permit remote real time monitoring to elucidate, diagnose and effect corrective action of equipment issues before they become a catastrophic problem, thus saving massive amounts of time and money.					

(*Continued*)

Table 6.8 Continued

Industrial Internet Applications	Devices					
	Switch	Varactor	Resonator	Accelerometer	Vibration Sensor	Energy Harvester
	X	X	X	X	X	X
Manufacturing Equipment Monitoring	In a factory floor, with a copious number of machines, wireless remote monitoring is crucial to maintain overall control of an efficient operation. Endowing these machines with "intelligence" facilitates data collection, reporting and analysis, providing insights into the overall equipment effectiveness and the anticipation/timeliness of machine maintenance. Also, the addition of wireless connectivity in an *ad hoc* fashion endows the factory with flexibility for quick reconfiguration if new machines are brought in line. Typical parameters of interest include vibration and temperature monitoring.					
	X	X	X	X	X	X
Asset Tracking and Smart Logistics	The field of logistics has grown tremendously in the last few years, due to the proliferation of large companies engaged in asset distribution. In particular, it has become imperative to have real time information on the location of railway cars, truck trailers, and containers containing valuable assets as they move from point to point. In the context of the Supply Chain, it is imperative to keep track of, and accurately record the many check-in and check-out events involved. The IoT can be exploited to enable the distributed tracking of the various elements in question via low power, cost effective, compact wireless nodes to relay real time information to a logistics control center.					

Source: Ref. [137–143].

Furthermore, with the imminence of self-driving cars, IoT connectivity will become mandatory for ubiquitous car monitoring.

6.3.7 NEMX in Connected Health IoT Applications

The connected health paradigm enabled by the IoT promises to extend the reach of healthcare professionals, while reducing the cost of healthcare services [134]. By enabling the performance of remote patient monitoring at their homes, instruments connected to the IoT will monitor and relay patient's status to their doctors and nurses. Table 6.10 lists examples of applications.

6.3.8 NEMX in Smart Retail IoT Applications

There are many motivations for exploiting IoT in the area of Smart Retail [135]. In general, all areas of retail stand to benefit from the real-time monitoring enabled by IoT, for instance, capturing traffic patterns in a store,

Table 6.9 Opportunities for MEMS/NEMS in Connected Car

Connected Car IoT Applications	Devices					
	Switch	Varactor	Resonator	Accelerometer	Vibration Sensor	Energy Harvester
Connected Vehicle Sensors	X	X	X	X		
	The connected car will be equipped with sensors to enable a plethora of monitoring activities such as effecting preventive maintenance and real-time vehicle diagnostics. In this context, for instance, a monitoring system can collect data from the motor starter, the fuel pump, and the battery. Then, via the connected car, this data can be transmitted to a cloud server which, upon analysis, can predict potential maintenance issues, subsequently delivering to the driver suggestions on how to fix these. Other sensors that are already beginning to appear in the market include those precluding collisions and allowing hands-free driving.					
Vehicle Usage Analytics	X	X	X	X	X	X
	This includes various types of analytics such as monitoring the real time driving patterns and road behavior of all individuals driving a certain kind of vehicle to assess driving behavior in relation to positive or negative outcomes.					
Vehicle Location Tracking and Scheduling Solutions	X	X	X	X	X	X
	With the explosion of transportation services, both of people and goods, it is imperative to keep track of vehicle location to enhance dispatch schedules and increase efficiency. This is an area of great importance enabled by vehicle connectivity to the Internet.					

Source: Ref. [144–145].

tracking inventory, and enabling automatic check out. Judicious exploitation of this technology is bound to increase business efficiency and profits. Table 6.11 lists examples of applications.

6.3.9 NEMX in Smart Supply Chain IoT Applications

The Supply Chain refers to the procession of good as they travel from their source to their destination. In the Smart Supply Chain, it is possible to track at all times the location of these goods and the speed at which they displace, enabling the optimization of the distribution route to achieve fastest, reliable, and cost effective on time delivery. Table 6.12 lists examples of Smart Supply Chain applications in the context of the IoT [152].

6.3.10 NEMX in Smart Farming IoT Applications

The idea of exploiting IoT for Smart Farming derives from a farmer's desire to gain better control over the process of raising livestock and growing crops,

Table 6.10 Opportunities for MEMS/NEMS in Connected Health

<table>
<tr><th rowspan="2">Connected Health Internet Applications</th><th colspan="6">Devices</th></tr>
<tr><th>Switch</th><th>Varactor</th><th>Resonator</th><th>Accelerometer</th><th>Vibration Sensor</th><th>Energy Harvester</th></tr>
<tr><td rowspan="2">Adjustable Patient Monitoring</td><td>X</td><td>X</td><td>X</td><td></td><td></td><td></td></tr>
<tr><td colspan="6">Due to an increasingly price-conscious consumer, health care providers must innovate to continue to become profitable. In this context, remote patient monitoring is essential, as it would allow healthcare professionals to service patients at their homes. An area of concern, that must be addressed, is that of the security concerns surrounding protection of patient data as it is transmitted or stored.</td></tr>
<tr><td rowspan="2">Enhanced Drug Management</td><td>X</td><td>X</td><td>X</td><td>X</td><td>X</td><td>X</td></tr>
<tr><td colspan="6">An issue that permeates the regular intake of drugs to obtain desired relief and effectiveness, is their timely administration. An important application of the IoT is the possibility of producing edible “smart” pills that will aid the monitoring of health issues and medication controls.</td></tr>
<tr><td rowspan="2">Early Intervention</td><td>X</td><td>X</td><td>X</td><td>X</td><td>X</td><td>X</td></tr>
<tr><td colspan="6">The possibility of IoT-driven remote monitoring of people will result in detecting a variety of life-threatening events such as the detection of a fall for a senior living alone, the drop in blood pressure, blood sugar levels, etc. IoT connectivity will immediately alert family members or emergency responders for timely intervention.</td></tr>
</table>

Source: Ref. [149].

making it more predictable and improving its efficiency [153]. Table 6.13 lists examples of Smart Farming applications in the context of the IoT [153].

6.4 Applications in Wireless Sensor Networks

As indicated in our previous exposition, the power consumption of the individual nodes integrating an IoT network is of prime importance, as it may directly determine, for example, in the case of battery-powered nodes, the degree of autonomy/life attainable by such a node. Much effort, therefore, has been aimed at reducing radio power dissipation, even as the utilized data rates set by communications protocols such as Bluetooth remains at a few hundred kb/s to a few Mb/s while consuming around a few mW power [154]. It appears, however, that exploiting clever circuit design techniques as a means to reducing radio power consumption has reached its point of diminishing returns, thus, new approaches are under investigation. One such approach is referred to as “duty cycling,” which aims at restricting the episodes of radio transmission to very short bursts of data at high rates while remaining asleep the rest of the time [154].

Table 6.11 Opportunities for MEMS/NEMS in Smart Retail

Smart Retail Applications	Devices					
	Switch	Varactor	Resonator	Accelerometer	Vibration Sensor	Energy Harvester
Foot-Traffic Monitoring	X	X	X	X	X	
	A key to increasing sales in a "brick and mortar" store is to place articles for sale where they can be found. In a smart store, an assessment of the areas where customers tend to spent most of their time may be monitored, utilizing foot-traffic monitoring tools such as video cameras, in real time to optimize store layouts.					
Monitor the Temperature Fluctuations to Ensure Food Safety	X	X	X	X	X	X
	As is well known, modern refrigerators are tasked with preserving food in conditions that are safe for consumption, and one of the key environmental parameters to achieve this is their temperature. Indeed, bacteria grow or reproduce rather slowly when under low or freezing temperatures. In particular, in the context of food stores, it is imperative to preserve food at the right temperature over long periods of time, so it is safe and ready for consumption when purchased. The ability of the IoT to link distributed sensors can be exploited to monitor and control the temperature of food whether in a truck, while being transported or in the store.					
Demand Alert Warehouses	X	X	X			
	With the proliferation of online shopping, fast and automatic fulfillment of purchased goods is becoming more and more desirable; a short delivery time is crucial or the customer may change his/her mind. In demand-alert warehouses, the IoT-connected RFID-enabled goods can enable the real time monitoring of sales. This, in turn, allows real time inventory management and optimization.					
Smart Fulfillment	X	X	X			
	To optimize delivery time, the IoT can play a crucial role to distribute merchandise. In particular, IoT can be exploited to optimize the tracking, efficient transport and routing of goods. This activity may involve the merging of wireless connectivity with the global positioning system (GPS).					
Automated Checkout	X	X	X			
	Enabling fast and automated check-out after the customer has chosen his/her merchandise is a key activity for both customer and store. The IoT may enable the optimization and, thus, increased efficiency of the check-out operation by allowing retailers to make the reading of tags automatic, thus eliminating checkout lines.					

Source: Ref. [150, 151].

A major problem with duty cycling, however, pointed out by Thirunarayanan, Heragu, Ruffieux, and Enz (THRE) [154], is the large amount of energy spent (wasted) during the turn-on (start-up) and turn-off phases of each burst, which is in fact comparable to the energy used to communicate. In searching for the source of this energy inefficiency, they

Table 6.12 Opportunities for MEMS/NEMS in Smart Supply Chain

<table>
<tr><th rowspan="2">Smart Retail Applications</th><th colspan="6">Devices</th></tr>
<tr><th>Switch</th><th>Varactor</th><th>Resonator</th><th>Accelerometer</th><th>Vibration Sensor</th><th>Energy Harvester</th></tr>
<tr><td rowspan="2">Authenticate the Location of Goods at Any Time</td><td>X</td><td>X</td><td>X</td><td></td><td></td><td></td></tr>
<tr><td colspan="6">The embedding of IoT nodes into products, including storage containers and raw materials, automatically endows them with wireless connectivity, which allows the real time tracking of their location.</td></tr>
<tr><td rowspan="2">Monitor Storage Conditions of Raw Materials and Products</td><td>X</td><td>X</td><td>X</td><td></td><td></td><td>X</td></tr>
<tr><td colspan="6">Certain goods such as food and chemicals must be kept under specific environmental conditions, in particular, temperature, humidity, light, etc. In this context, IoT-connected sensors will be crucial to emit alert signals every time certain thresholds are violated, thus enabling the preservation of the quality of said goods.</td></tr>
<tr><td rowspan="2">Streamline the Problematic Movement of Goods</td><td>X</td><td>X</td><td>X</td><td></td><td></td><td>X</td></tr>
<tr><td colspan="6">In the chaotic world of merchandise distribution, the IoT can enable the ability track the location of goods that are in-transit. This connectivity can help in devising strategies to track and route the movement of goods so they arrive at their destination on time.</td></tr>
<tr><td rowspan="2">Locate Goods in Storage</td><td>X</td><td>X</td><td>X</td><td></td><td></td><td></td></tr>
<tr><td colspan="6">The utilization of smart tags, embodying IoT nodes, on goods, can facilitate their identification and location while in a large warehouse or distribution center.</td></tr>
<tr><td rowspan="2">Administer Goods Immediately Upon Receipt</td><td>X</td><td>X</td><td>X</td><td></td><td></td><td>X</td></tr>
<tr><td colspan="6">The same smart tags indicated above may be utilized to also track and ascertain the exact time of arrival of goods. This will facilitate the identification of contract compliance terms that may then cause the initiation of contractual payments.</td></tr>
</table>

Source: Ref. [152].

identified the main source of the turn-on/off energy drain as due to the long start-up of the loop-based frequency synthesizer, in particular, in the crystal oscillator frequency reference. The solution they proposed was the utilization of MEMS Film Bulk Acoustic Wave Resonators (FBARs) [81] which, due to their high quality factor (Q), would enable reducing the radio turn-on/off times from the order of milliseconds to a few microseconds [154].

6.4.1 NEMX-Based Radios for the IoT

The solution proposed by THRE [154] is exemplified by a MEMS-based radio transceiver, in which high-Q MEMS FBARs are used to implement a digitally-controlled oscillator circuit (DCO) (Figure 6.8).

Table 6.13 Opportunities for MEMS/NEMS in Smart Farming IoT

Smart Retail Applications	Devices					
	Switch	Varactor	Resonator	Accelerometer	Vibration Sensor	Energy Harvester
Monitoring of Climate Conditions	In this application, weather stations that combine a number of smart farming sensors to enable the collection and mapping of climate and environmental data to help, for example, choosing the appropriate crops and take the required actions to improve their growth.					
	X	X	X	X	X	X
Greenhouse Automation	In this application, weather stations are employed to automatically adjust the conditions to match desired parameters.					
	X	X	X	X	X	X
Crop Management	In this application, data collection that is specific to crop farming, in particular, temperature, precipitation, and leaf water potential and overall crop health is taken. The farmer can monitor crop growth and any anomalies to enable the prevention of any diseases or infestations that can harm yield.					
	X	X	X	X	X	X
Cattle Monitoring and Management	In this application, sensors are attached to cattle on a farm to monitor their health and log performance. This helps the farmer to develop insights on animal temperature, health, activity, and nutrition on each individual cow as well as aggregate information about the herd.					
	X	X	X	X	X	X
End-to-End Farm Management Systems	In this application, a global view of the whole farming operation may be exploited to enhance and manage the overall farm productivity. In particular, devices and sensors may be deployed on the premises that contain powerful analytical capabilities and built-in accounting/reporting features. As a result, the remote farm monitoring capabilities obtained enable the streamlining of most of the business operations.					

Source: Ref. [153].

The FBAR-based DCO is used as the key building block for the local oscillator in the receiver [81]. Also, because of its high Q (in the thousands), the FBAR DCO is responsible for enabling a low phase noise signal that eliminates the need for a phase-locked loop (PLL) to address multiple channels. In addition, the FBAR DCO is utilized in the generation of high spectral purity clocks for general usage throughout the receiver. Finally, the high-Q FBAR is also employed in the RF front-end to realize highly selective filtering functions such as the rejection of in-band interferers.

On the other hand, in the transmitter side, the FBAR DCO is employed to produce the reference local oscillator frequency (LO).

The positive impact of using the high-Q FBAR-based DCO was measured as a transmitter exhibiting a wake-up time of 5 μs, representing an improvement of approximately 200 times reduction compared to that of a

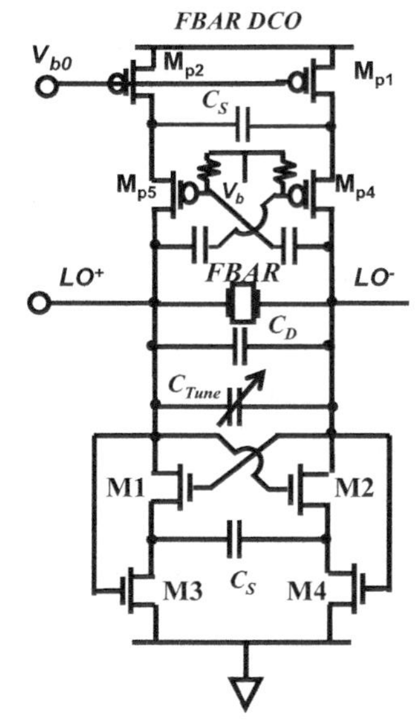

Figure 6.8 High Q FBAR oscillator.

Source: Ref. [154].

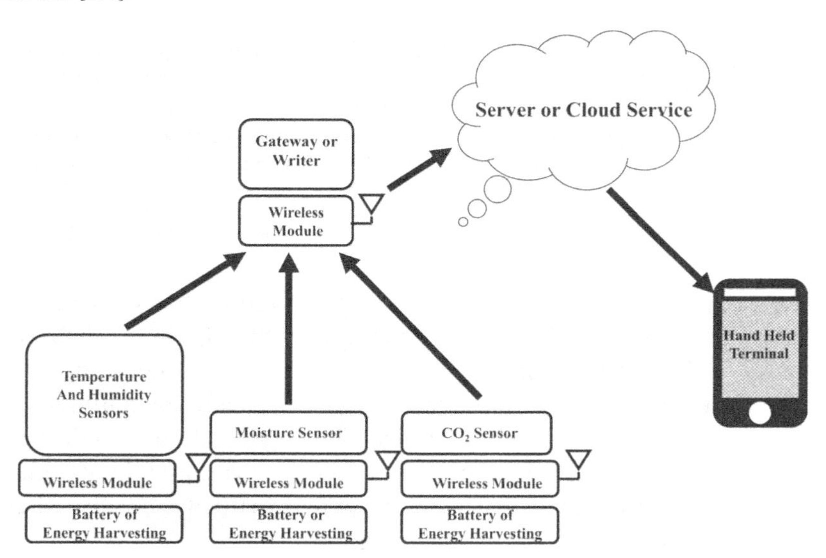

Figure 6.9 Wireless Sensor Network deployment for the agricultural applications.

Source: Ref. [155].

crystal oscillator and leading to a reduction in energy dissipated by a factor 34 compared to that consumed by conventional radios during their start-up phase [154].

6.4.2 Agricultural Applications

One of the key motivations for deploying IoT nodes in a farm is the so-called precision agriculture [155]. This refers to the precise comprehension, from a scientific point of view, of the relation between crop conditions and their irrigation needs, and their correct application during both the harvesting and sowing seasons. The ultimate outcome of precision agriculture is to increase productivity and minimize the unintended effects of equipment failure, an adverse environment and wildlife [155].

A prototypical wireless sensor network would include a number of standalone nodes (Figure 6.9). These nodes would contain an energy source such as a battery or an energy harvesting module, a wireless (transceiver) module and a variety of physical sensors, for example, soil moisture, humidity, pressure, wetness and temperature sensors. The system would also include a base section to serve as the gateway for communication between the end users and the nodes or between nodes [155].

6.5 5G: Systems [156]

It should be clear by now, that the Internet of Things will ultimately encompass a network of billions of nodes connected wirelessly that may ultimately be spread around the world and, thus, have a global reach. Obviously, since the frequency spectrum is a limited resource, new frequency bands will have to be opened up to accommodate the explosion in data rates that will ensue. In fact, Cisco, in its annual visual network index (VNI) reports, has provided decisive evidence to this effect [156]. In particular, in Cisco's 2014 report [156] it is forecasted that, since an incremental approach will not be enough to meeting the demands that networks will face by 2020, a more radical approach to high-capacity wireless communications will be necessary.

Indeed, according to Andrews et al. [156], the progression in the amount of IP data handled by wireless networks will have increased from over 190 exabytes[1] by 2018 and to well over 500 exabytes by 2020. The wireless ecosystem behind this large volume of data will include, not only video, but

[1]One exabyte equals 1 billion gigabytes.

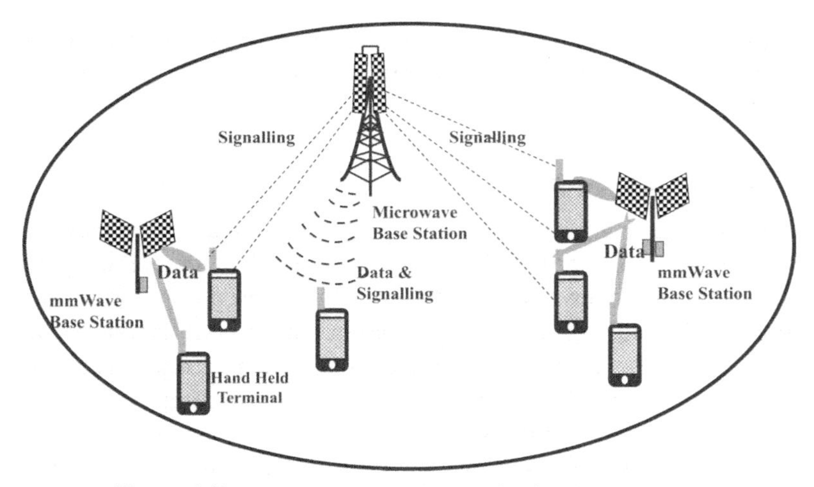

Figure 6.10 mmWave-enabled network with phantom cells.

Source: Ref. [156].

the number of connected IoT devices, which is expected to reach possibly into the hundreds of billions, hence, the motivation for 5G.

5G, the fifth generation of wireless communications standards [157], is projected as being enabled by the exploitation of three technological areas to enhance network capacity, namely [156]: 1) Ultra-densification, which refers to improving the spectral density by placing more active nodes per unit area and Hz; 2) mmWave, which refers to increasing the available communications bandwidth by moving into the mmWave spectrum to have more Hz available; and 3) Massive Multiple-Input Multiple-Output (MIMO), which refers to increasing spectral efficiency through advances in MIMO receive/transmit (antenna) nodes to achieve more bits/s/Hz per node.

In this section, focus is placed on mmWaves. This is the range of frequencies encompassing 30–300 GHz, with corresponding wavelengths in the 1–10 mm range; NEMX is expected to be an enabling technology for mmWave nodes. Another frequency that tends to be lumped with mmWaves, when they are considered, is the 20–30 GHz range, which may also be exploited.

A number of reasons have been voiced against the utilization of mmWaves for cell phone network-like wireless communications. These include:

(1) Propagation Issues, which refers to the fact that, as the frequency increases, the antenna size decreases and its aperture scales as $\lambda^2/4\pi$,

where $\lambda = c/f_c$ is the wavelength, c is the speed of light and f_c is the carrier frequency, so that the free-space path loss [157] between a transmit and a receive antenna grows as f^2. This concern has been ameliorated, however, with the realization that by using arrays of antennas, it can be overcome.

(2) Blocking, which refers to the fact that, since at mmWave signal propagation undergoes reduced diffraction and more specular propagation than their microwave counterparts, it is more sensitive to blockages. This is responsible for an additional loss of 15–40 dB per decade compared to the free-space path loss value of 20 dB/decade as the transmit/receive antenna distance increases, and is responsible for the rapid link transition from usable to unusable, which cannot be circumvented with standard smalls scale diversity countermeasures.

(3) Atmospheric and Rain Absorption, which refers to the fact that the absorption due to air and rain is not negligible, particularly due to the 15 dB/km oxygen absorption within the 60 GHz band, but negligible in the setting of urban cellular deployments envisioned, where base station (BS) spacings would not exceed 200 m. In summary, it has been determined that by using large antenna arrays to steer the energy and collect it in a coherent fashion, the problem of propagation losses at mmWave frequencies is surmountable, but it requires the production of narrow beams.

(4) Large Arrays, Narrow Beams, which refers to the fact that since, in contradistinction to traditional wireless systems at lower frequencies, new territory is encountered when it comes to the design of wireless systems predicated upon the use of narrow beams. This derives, in particular, from the abruptness of the interference behavior of highly directional, pencil beams, which augments the sensitivity to beam misalignment. Thus, the interference adopts an on/off behavior where the majority of the beams do not interfere, but rather, events of strong interference are experienced intermittently.

(5) Link Acquisition, which refers to the fact that, due the narrow beams characteristics at mmWaves, there is a concomitant difficulty to establish links between base stations (BSs) and users; this creates a problem at both the time of initial access and handoff.

To overcome the above problems, the integration of microwave and mmWave frequencies has been considered (Figure 6.10). This scenario posits the utilization of mmWave frequencies for payload data transmission from small-cell BSs, together with the utilization of microwave frequencies in macro BSs for beam control. This concept affords the natural integration of IoT nodes, which then could take advantage of mmWaves when properly designed to operate at these frequencies. Considerations of realizing IoT nodes capable of exploiting mmWaves dictates that changes in IoT node transceivers similar to those required in a mmWave cellular network be made. In particular, this will call for new transceiver architectures, power-efficient analog-to-digital (A/D) and digital-to-analog (D/A) converters operating on large bandwidths. One potential advantage of IoT nodes, that makes mmWave operation closer to feasibility, is their close proximity, which would probably bypass the path loss issue, although the beam pointing problem would still remain due to their smaller size. This area, therefore, is full of interesting challenges.

6.6 5G: Technologies [158]

The desire to exploit the advantages afforded by mmWaves/5G in the context of the IoT has elicited a number of new concepts to deal with the limiting issues presented above, which have been discussed by Choudhury [158].

6.6.1 Device-to-Device Communications

A promising technology in the context of mmWaves is that of Device-to-Device (D2D) communications. The idea behind this technique is to avoid the base station as a bottleneck between source and destination. Thus, communication traffic would go directly from one device to nearby devices through the establishment of local links. In particular, D2D communication has the potential to reduce latency and power consumption, increasing peak data rates, and enhancing spectrum reuse since many D2D links could be designed so that they would share the same bandwidth, resulting in an increase of spectral reuse per node.

6.6.2 Simultaneous Transmission/Reception

In the simultaneous transmission and reception (STR) technique, transmission and reception are effected during the same time and through the same frequencies, so as to enable a higher spectral efficiency. Thus, a doubling of the spectral efficiency is immediately achieved in D2D links. In addition, STR makes easier the discovery of neighboring device in the context of

D2D communication networks because it becomes possible to monitor uplink signals from adjacent nodes without momentarily suspending transmission.

6.6.3 mmWave/5G Frequencies for IoT

Many frequency bands appear to be promising for IoT. These include, 10–15 GHz, 28–30 GHz, 38–40 GHz, 57–66 GHz, 71–76 GHz, 81–86 GHz, and 92–95 GHz [156, 158, 159]. As the frequency increases beyond a few GHz, that is, the usual 6 GHz for cell phones, the relevance of NEMX becomes more and more apparent, due to its ability to enable passive devices, for example, switches and varactors, with superior insertion loss and isolation compared to the currently dominant semiconductor devices. These NEMX devices, in turn, will enable superior circuits such as phase shifters for the realization of the agile antenna arrays [18, 81, 158] needed to create pencil beams for establishing links between IoT nodes.

6.7 Summary

In this chapter, we have dealt with the NEMX applications in the IoT era. In particular, after an introduction of the fundamentals of IoT networks and nodes, we explored how the NEMX components are encroaching in a variety of IoT applications. We saw that, while the applications of the IoT are virtually unlimited, it is useful in studying them, to organize them in terms of the following realms: (i) Smart Home; (ii) Wearables; (iii) Smart City; (iv) Smart Grid; (v) Industrial Internet; (vi) Connected Car; (vii) Connected Health; (viii) Smart Retail; (ix) Smart Supply Chain; (x) Smart Farming. The popularity of these applications appeared ranked in Figure 6.6. Next we placed NEMX in the context of their opportunities for insertion into these applications, and finally, we addressed the subject of the IoT in the context of emerging mmWaves/5G (fifth generation wireless networks) technology.

Appendix A: MEMS Fabrication Techniques Fundamentals

A.1 Introduction

This appendix introduces the fundamental fabrication techniques employed in the fabrication of three-dimensional microelectromechanical systems. In doing so, we assume the reader has a basic knowledge of the conventional integrated circuit (IC) fabrication process, although a brief review is presented to motivate an appreciation for MEMS fabrication technologies.

A.2 The Conventional IC Fabrication Process

The traditional IC fabrication process utilizes photolithography and chemical etching (Figure A.1) [161]:

(1) The process begins by covering the wafer with a thin-film material barrier (typically SiO_2 in silicon technology or silicon nitride in gallium arsenide technology) on which a pattern of holes is to be defined.
(2) A light-sensitive material, called photoresist, is deposited on the thin-film material coating the wafer surface.
(3) A square glass plate, called photomask, containing a patterned emulsion or a metal film on one side, is placed over the wafer such that upon illumination with high-intensity ultra violet (UV) light, those areas with oxide to be removed are exposed to UV light.
(4) A process very similar to that utilized in developing ordinary photographic film is used to develop the image impressed on the photoresist.

The resulting image defined on the wafer surface depends on the nature of the photoresist.

- If the photoresist that has been exposed to UV light is washed away, leaving bare SiO_2 in the exposed area, it is a positive resist, and the

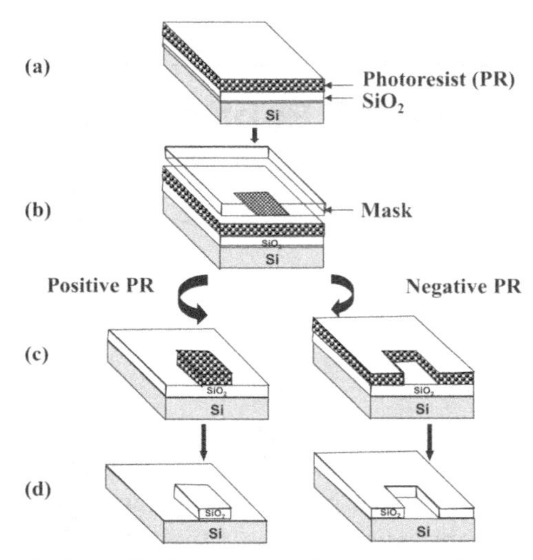

Figure A.1 Resist and silicon dioxide patterns following photolithography with positive and negative resists.

Source: Ref. [161].

mask contains a copy of the pattern, which will remain in the surface of the wafer.

- A negative resist remains on the surface wherever it is exposed.

This is the essence of a conventional IC fabrication process; it permits the transfer of a circuit layout pattern onto a two-dimensional wafer surface.

Traditional IC processes are not empowered, however, to realize three-dimensional structures, that is, form three-dimensional microelectromechanical devices. This is the realm of two processes devised for this purpose, namely bulk micromachining and surface micromachining [162].

A.3 Bulk Micromachining

Bulk micromachining undertakes the fabrication of mechanical structures in the bulk of a wafer, in contradistinction to fabricating it on its surface. This involves the selective removal of some parts of the wafer/substrate material to form trenches and then bond wafers together, via a mediating bonding layer, to create cavities (Figure A.2).

The technology is currently considered mature technology, as it was originally developed for the production of silicon pressure sensors in the late 1950s. New techniques, aimed at producing three-dimensional

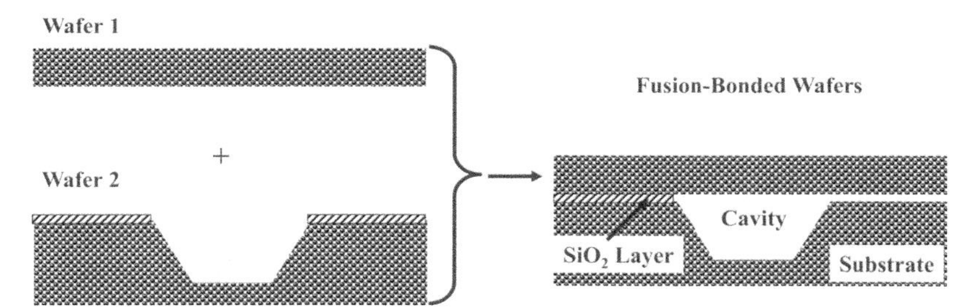

Figure A.2 Formation of cavity using bulk micromachining (LEFT) and wafer bonding (RIGHT).

micromechanical structures, however, have emerged, namely surface micromachining, silicon fusion bonding, and a process called the LIGA, a German acronym consisting of the letters LI (Roentgen LIthographie meaning X ray lithography), G (Galvanik meaning electrode position), and A (Abformung meaning molding) process. These technologies are based on the use of photolithography, thin-film deposition, and chemical etching, which are compatible with standard IC batch processing; therefore, they exhibit great potential for enabling novel complex systems.

Bulk micromachining is enabled by the process of chemical etching. In this process, by combining highly directional (anisotropic) etchants, with non-directional (isotropic) etchants, together with the wafer's crystallographical orientation, the etching rates are manipulated to define a wide variety of mechanical structures within the confines of the wafer bulk. Furthermore, by creating contours of heavily doped regions, which etch more slowly, and p–n junctions, which stop the etching process altogether, it is possible to form deep cavities. Deep cavities are fundamental to the engineering of many devices, for example, diaphragms for pressure sensors and low loss planar inductors.

Despite its maturity, bulk micromachining traditionally has some fundamental limitations. For example, the fact that the wafer's crystallographic planes determine the maximum obtainable aspect ratios poses a restriction to the attainable device geometry, namely it results in relatively large sizes compared with other micromachining techniques [163].

A.4 Surface Micromachining

In surface micromachining, thin-film material layers are deposited and patterned on a wafer/substrate. Thin-film material, which is deposited wherever either an open area or a free-standing mechanical structure is desired, is called *sacrificial* material. On the other hand, the material out of which the free-standing structures are to be made is called *structural* material. In this process, to define a given surface micromachined structure, therefore, a sequence of wet etching, dry etching, and thin-film deposition steps must be discerned. Early demonstrations of the potential of surface micromachining were advanced in the 1960s and 1970s by scientists at Westinghouse Electric Corp., Pittsburgh, and at IBM Corporation. At Westinghouse, developments included micromechanical switches and electronic filters that use mechanically resonant thin-film metal structures, as well as advanced light-modulator arrays. At IBM, on the other hand, developments centered on the application of surface micromachining principles to displays, electrostatically actuated mechanical switches, and sensors, in which thin-film oxide structures were integrated with microelectronics [163].

A clear point of departure for the development of this technology was the 1967 paper "The Resonant Gate Transistor" [164], which described the use of sacrificial material to release the gate of a field-effect transistor. This work demonstrated the ability of silicon fabrication techniques to free mechanical systems from a silicon substrate. The next key development on surface micromachining was the use of polysilicon as the structural material, together with silicon dioxide as the sacrificial material, and hydrofluoric acid (HF) to etch silicon dioxide [165].

The next important achievement in the development of surface micromachining technology was its adoption of structural polysilicon and sacrificial silicon dioxide to fabricate free-*moving* mechanical gears, springs, and sliding structures [165, 166]. Since systems applications require that sensors and actuators interface with electronic circuitry, attention turned to the simultaneous fabrication of micromechanical devices with integrated circuits [167, 168]. Initial devices included polysilicon microbridges [167] and resonant microstructures, which were fabricated together with conventional complementary metal oxide semiconductor (CMOS) and N-type MOS processes, respectively. In this context, thin films of polysilicon, grown and deposited silicon dioxide, nitride materials, and photoresist, usually provide sensing elements and electrical interconnections, as well as structural, mask, and sacrificial layers. Released mechanical layers have been made with silicon dioxide, aluminum, polyimide, polycrystalline silicon, tungsten, and single-crystal silicon. Figure A.3 shows the key steps involved in surface micromachining.

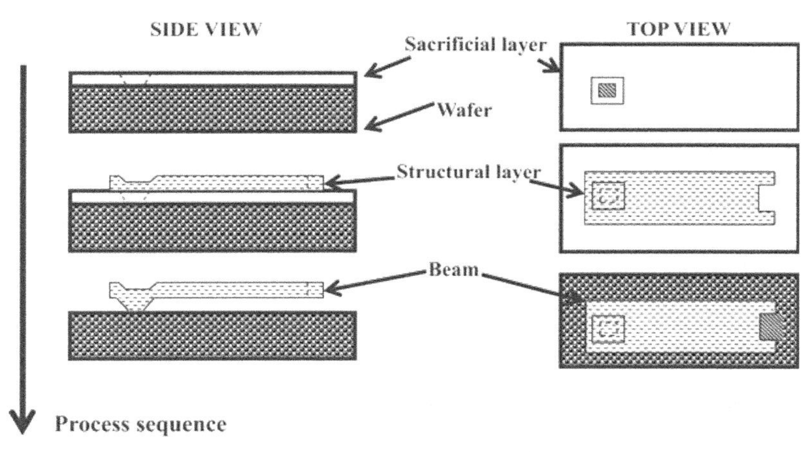

Figure A.3 In the surface micromachining process, a sacrificial layer is grown or deposited and patterned, and then removed wherever the mechanical structure is to be attached to the substrate. Then, the mechanical layer is deposited and patterned. Finally, the sacrificial layer is etched away to release the mechanical structure.

A number of alternatives to silicon as a substrate material exist. One such alternative came about with the intent of applying surface micromachining techniques to incorporate microwave MEM devices in monolithic microwave-integrated circuits (MMICs). This entailed exploitation of the semi-insulating low-loss properties of gallium arsenide (GaAs) wafers, and culminated in the MIMAC process [170]. In a more recent instance, a surface micromachining process that exploits the low-loss properties alumina substrates and the inexpensive nature of alumina-based microwave-integrated circuits (MICs) was also developed [171].

A.5 Materials Systems

An examination of the MEMS fabrication literature [172–175] reveals the emergence of a number of structural/sacrificial/etchant material systems used together. A partial list of these is presented in Table A.1.

Examination of Table A.1 reveals that, for example, an aluminum structure may be deposited over photoresist and then set free by effecting its release with oxygen plasma; that a polysilicon structure may be deposited over silicon dioxide and set free by release with HF; and that a silicon dioxide structure may be deposited over polysilicon and set free by release with xenon-difluoride.

Table A.1 Structural/sacrificial/etchant material systems

Structural Material	Sacrificial Material	Etchant
Aluminum	Single-crystal silicon	EDP, TMAH, XeF_2
Aluminum	Photoresist	Oxygen plasma
Copper or nickel	Chrome	HF
Polyimide	Aluminum	Al etch (phosphoric, acetic, nitric acid)
Polysilicon	Silicon dioxide	HF
Photoresist	Aluminum	Al etch (phosphoric, acetic, nitric acid)
Aluminum	Single-crystal silicon	EDP, TMAH, XeF_2

A.6 Summary

In this appendix, we have presented the fundamentals of MEMS fabrication technology, in particular bulk micromachining, surface micromachining, together with materials systems they employ. A more detailed and in-depth study of MEMS/NEMS fabrication technology may be found in a number of sources in the literature, for example, [5, 6, 10, 72].

Appendix B: Emerging Fabrication Technologies for the IoT: Flexible Substrates and Printed Electronics

B.1 Flexible Substrates [176]

As discussed in Chapter 1, the Internet of Things (IoT) will be predicated upon the wireless networking of billions of nodes. These nodes contain sensors, actuators, wireless transceivers, energy storage and harvesting devices, power management, and a microcontroller brain. Clearly, such an enormous number of devices can only materialize commercially if cost-effective ways for fabricating them are brought to bear. In this context, a number of technologies have been under development to fabricate virtually inexpensive electronics as it is necessary to build IoT nodes, and the state of development of these is rather timely in terms of their degree of readiness for application in the IoT.

Flexible substrates is one such technology that enables the fabrication of healthcare, environmental monitoring, displays and human–machine interactivity, energy conversion, management and storage, and communication and wireless networks (Figure B.1) [176]. The technology, which is derived from advances in thin-film materials and devices, is fueling many of the developments in the field of *flexible electronics*, so that electronic components predicated upon the integration of both passive components such as resistors, capacitors and inductors, and active components such as diodes and transistors are available for utilization in a plethora of digital and analog circuits, as well as for detection and energy generation.

The word "flexible" arises from the fact that the substrates in question, on which devices are fabricated via thin-film deposition techniques, are plastic in nature. As a result, they may be used to make devices that can be attached onto non-planar surfaces, such as airplanes. Table B.1 lists a number of such plastic materials. The realization of a number of devices, with potential pertinence to IoT applications, is presented subsequently.

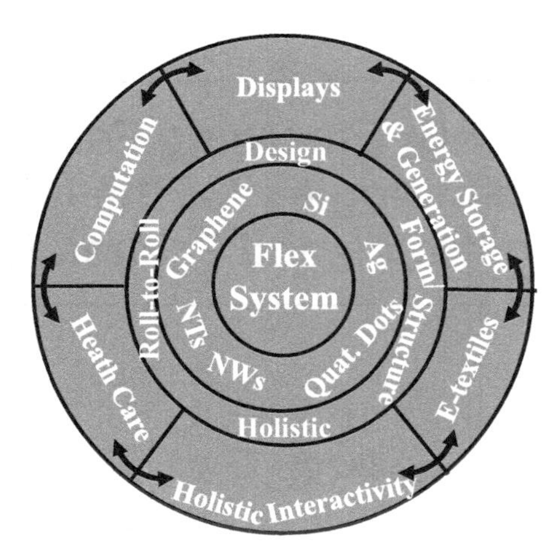

Figure B.1 Future flexible electronics systems and the key relevant sectors, the underlying materials—such as the industry pervading, historically relevant and standard aluminum, silicon, germanium, and silver, as well as more exotic low-dimensional materials including nanowires, quantum dots, and nanotubes—all of which will be necessary to facilitate the development and the exploitation of disruptive applications in the fields of human interactivity, computation, displays, energy generation, and storage as well as electronic textiles.

Source: Ref. [176].

Table B.1 Comparison of possible plastic substrate for thin-film deposition

Max. Deposition (Temp. °C)	Material	Properties
250	Polyimide (Kapton)	Orange color; high thermal expansion coefficient; good chemical resistance; expensive; high moisture absorption
240	Polyetherketone (PEEK)	Clear, good dimensional stability; poor solvent resistance; expensive; moderate moisture absorption
190	Polyethersulphone (PES)	Clear; good dimensional stability; poor solvent resistance; expensive; moderate moisture absorption
180	Polyetherimide (PEI)	Strong; brittle; hazy color; expensive
160	Polyethlene naphtalate (PEN)	Clear; moderate CTE; good chemical resistance; inexpensive moderate moisture absorption
120	Polyethylene terephthalate (PET)	Clear; moderate CTE; good chemical resistance; inexpensive; moderate moisture absorption

CTE denotes the coefficient of thermal expansion.
T_{Max} denotes the maximum deposition temperature.
Source: Ref. [176].

B.1.1 Device Fabrication on Flexible Substrates

B.1.1.1 Thin-Film Transistors (TFTs)

The degrees of freedom afforded by thin-film processes and materials enable one to circumvent certain limitations germane to flexible substrates to produce high-performance devices, in the context of attaining a good compromise among device characteristics, cost, and stability. For example, while plastic substrates exhibit small electron mobilities, using short channel lengths may allow one to circumvent this limitation, thus rendering transistors that achieve, for example, high transconductance. Figure B.2 presents an

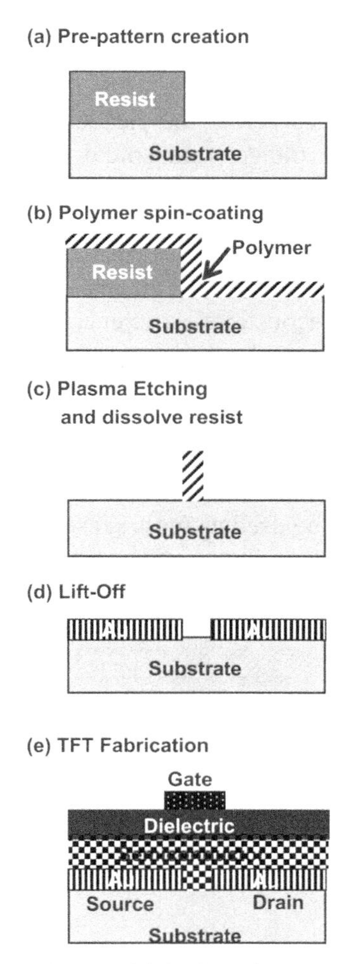

Figure B.2 Schematic illustrating the fabrication of short-channel transistors using spin-coating-induced edge effect.

approach to fabricate organic thin-film transistors (TFTs) uniformly over large areas [177]. The process exploits a novel edge effect that is induced by spin coating a polymer onto a pre-patterned structure. The method has yielded polymer TFTs, with channel widths as narrow as 400 nm, exhibiting a mobility of 5×10^{-3} cm^2/Vs and on/off current ratio of approximately 10^6 on poly (9, 9-dioctylfluorene-co-bithiophene) (F8T2).

B.1.2 Film Bulk Acoustic Wave Resonators (FBARs)

The film bulk acoustic wave resonator may be construed as consisting of a capacitor in which the dielectric is a piezoelectric material (Figure B.3) [81].

The piezoelectric effect, which the FBAR exploits, exhibits the following behavior: A force applied across a layer of piezoelectric material with piezoelectric coefficient d causes a voltage V to be induced across it. Conversely, when a voltage V is applied across the piezoelectric layer, it experiences a strain. Now, if the applied voltage is sinusoidal, then an oscillation is induced in which the voltage causes an acoustic wave to propagate between the electrodes, which in turn produces an electric field between the electrodes. This exchange between acoustic and electrical energy occurs at a certain rate, which means that the device behaves as an electric resonator, for example, an LC resonator. The rate of acoustic-electric energy exchange is the resonance frequency of the resonator and reaches several GHz. The interest on this FBAR device, as noted in Chapter 6, is that it exhibits a high-quality factor, namely of thousands; thus, it enables low-phase noise frequency oscillators and low insertion loss filters [81]. Furthermore, FBAR is small in size, is relatively inexpensive to manufacture, and is compatible with traditional CMOS. FBARs may also be used as mass sensors, in whose application, the physical adsorption of mass on its top plate causes mall loading, that is, slows down the resonance frequency and, thus, can be detected/sensed.

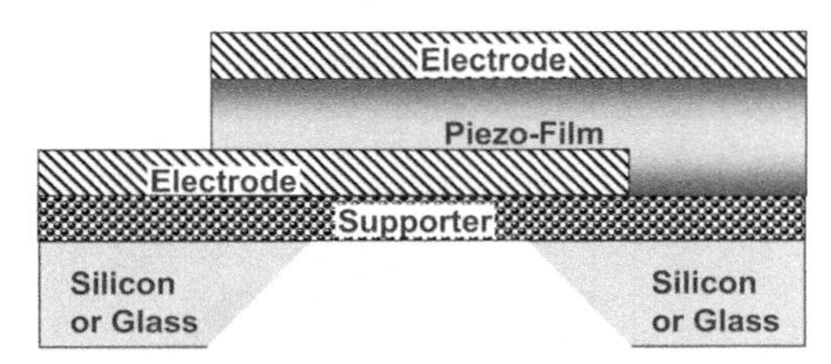

Figure B.3 Typical thin-film FBAR structure.

Source: Ref. [177].

B.2 Printed Electronics [178]

The field of printed electronics (PE) deals with the fabrication of devices using an inexpensive printing process. Devices, including antennas, sensors, and membrane switches, have already been abundantly demonstrated. One of the key motivations behind PE is the ability to print devices on flexible substrates, but using a printing machine. In fact, the combination PE with flexible polymers poses many advantages, when compared to fabrication on the traditional hard substrates, in particular higher contact area, the capability to fold/roll, and attaining lightweight.

The fundamental reasons that have resulted in the popularity and adoption of PE are its ability to produce easily fabricated devices that exhibit high performance and long-term reliability. The PE process involves the organized, patterned deposition on a substrate of a sequence of pastes, inks, or coatings that can derive from both organic and inorganic materials. Thus, it is possible to define metallic patterns using inorganic inks containing metallic nanoparticles such as copper, gold, silver, or aluminum that are dispersed onto a substrate. On the other hand, the utilization of organic inks, in particular those that derive from materials such as polymers (conductors, semiconductors, and dielectrics) may be employed in the fabrication of batteries, electromagnetic shields, capacitors, resistors and inductors, and sensors. If the ink derives from organic materials, like organic semiconductors, it may be used to fabricate the active layer of organic photo diodes (OPDs), organic light emitting diodes (OLEDs), organic field-effect transistors (OFETs), organic solar cells (OSC), and a variety of sensors.

A variety of printing electronics technologies are available (Figure B.4). A brief introduction to the various printing technologies is presented next.

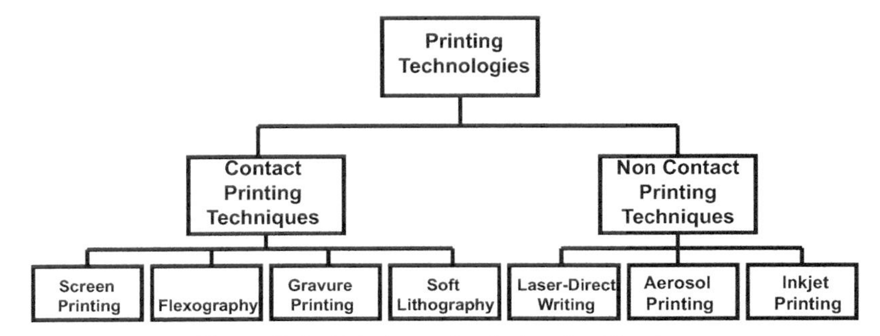

Figure B.4 Printing technologies classification.

Source: Ref. [178].

B.2.1 Printing Technologies

B.2.1.1 Contact Printing Techniques

The main contact printing technologies in use are as follows:

(1) *Screen printing*: In this technique, ink is first deposited on a stencil/screen, a sheet containing areas through which ink can transfer (apertures) and areas that block the ink [179]. Then, a blade or "squeegee" is moved across the screen to fill the apertures with ink, and the screen is pushed down and made to momentarily touch a substrate underneath. This causes the ink to wet the substrate and be pulled out of the mesh apertures as the screen is removed upwards after the blade has passed. By repeating this procedure with inks of different colors, a color image may be formed.

(2) *Flexography*: In this technique, printing is effected by utilizing a flexible photopolymer relief plate, wrapped around rotating cylinders, such that when it comes into contact with the receiving material, ink from the image defined by the relief transfers over onto it. It may be used on any type of substrate, including plastic, metallic films, cellophane, and paper [180].

(3) *Gravure printing*: In this technique, an image is etched with acid on the surface of a metal cylinder according to a pattern of cells. The cells, which are recessed into the etched cylinder, possess different depths. Then, ink filling the cells is transferred to a substrate. In this scheme, the deeper cells, containing more ink, produce a more intensive color than the shallower cells [181].

(4) *Soft lithography*: In this technique, an elastomeric stamp or mold, typically poly-dimethylsiloxane (PDMS) with patterned relief structures on its surface, is used to transfer patterns and structures with feature sizes ranging from 30 nm to 100 μm [72].

Table B.2 compares the properties of the various contact printing techniques.

B.2.1.2 Non-Contact Printing Techniques

The main non-contact printing technologies in use are as follows (Figure B.5):

(1) *Laser direct writing (LDW)*: In this technique, a computer-controlled laser beam is used to realize one-dimensional (1D), two-dimensional (2D), or three-dimensional (3D) structures. This is accomplished by

Table B.2 Comparison between main contact printing techniques

Printing Technique	Solution Types	Solution Viscosity (Pa.s)	Print Thickness (μm)	Resolution (μm)	Surface Tension (mN/m)
Screen printing	Water based, solvent based, UV or electron beam curable	0.1–10	0.02–100	30–100	38–47
Flexography	Water based, solvent based, UV curable	0.01–0.1	0.17–8	30–80	13.9–23
Gravure printing	Water based, solvent based, UV curable.	0.01–1.1	0.02–12	50–200	41–44
Soft lithography	Water based, solvent based, UV curable	~0.10	0.18–0.7	0.03–100	22–80

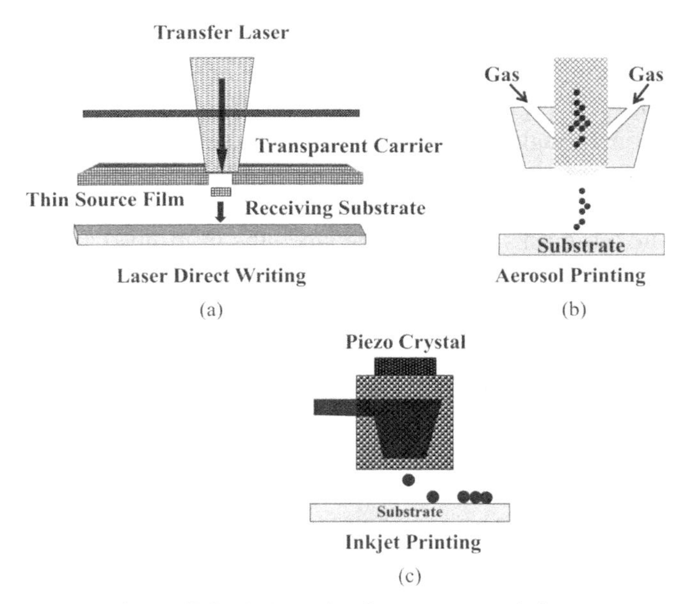

Figure B.5 Schematic of non-contact printing.

Source: Ref. [178].

inducing the deposition of metals, semiconductors, polymers and ceramics interposed between the laser and the substrate, according to the pattern described by the laser beam. The process does not utilize masks or physical contact between a tool or nozzle and the substrate material.

Through the control of the laser pulses, and the materials on which they impinge, the composition, structure, and properties of individual 3D volumes of materials are formed through length scales spanning from nanometers to millimeters.

(2) *Aerosol jet printing*: In this technique, the ink is held by an atomizer. Then, depending on the viscosity, it aerosolizes into liquid particles of diameter between 20 nm and 5 μm. The liquid consists of solutions and nanoparticle suspensions that may be metals, alloys, ceramics, polymers, adhesives, or biomaterials. During printing, the ink is transported into a deposition head by a nitrogen flow, while the aerosol is focused by a jet stream onto the substrate. As a low-temperature process, the technique is amenable for applying many materials and substrates. A high volume technique, it may be employed in the production designs that are both non-planar and complex such as displays, thin film transistors, TFT, and solar cells.

(3) *Inkjet printing*: In this technique, as the name implies, a pattern stored in memory in digital format is transferred onto a substrate via direct deposition through a print-head, much as when printing with a computer printer. Since the pattern printed is formed by controlling through which of the ejectors/pixels in the print-head the liquid/ink is ejected onto the substrate, no masks or contact between the print-head and the substrate are necessary. The ink may be in fluid or powder form and consist of proteins or minerals, conductive polymers, nanoparticles, or a wide variety of materials such as bioactive fluids. In operation, a piezoelectric material expands to push the precise amount of ink and together with the force of gravity and air resistance, the ink is expulsed onto specific positions on the substrate to create the printing patterns/images.

The technique can print onto a variety of textured substrates such as smooth or rough surfaces, for example, glass, plastic, paper, and textiles, with low consumption of ink materials. A number of applications have been realized via inkjet printing, including transducers, transistors, structural polymers and ceramics, biomimetic and biomedical materials, and printed scaffolds for growth of living tissues, as well as for 3D electric circuits, MEMS, and sensors.

Table B.3 compares the properties of the various non-contact printing techniques.

Table B.3 Comparison between the main non-contact printing techniques

Printing Technique	Solution Types	Solution Viscosity (Pa.s)	Print Thickness (μm)	Resolution (μm)	Surface Tension (mN/m)
Laser direct writing	Solid film (donor substrate)	–	>10	~0.7	–
Aerosol	Solutions and nanoparticle suspensions based on metals, alloys, ceramics, polymers, adhesives or biomaterials	0.001–1	>0.1	10–250	–
Inkjet	Water based, solvent based, UV curable	0.002–0.1	0.01–0.5	15–100	15–35

Source: Ref. [178].

B.3 Summary

As previously discussed, flexible electronics uses thin-film techniques to form devices on flexible substrates, while printed electronics uses some form of layered deposition technique to define patterns on virtually any type of substrate. Together, they provide a multitude of avenues to realize IoT nodes, including traditional passive and active planar devices, and also MEMS devices in a rapid and inexpensive fashion. Brought to full fruition, these techniques may become important tools to cost-effectively realize the various building blocks germane to IoT nodes, for example, sensors, actuators, wireless transceivers, energy storage and harvesting devices, power management, or microcontrollers.

References

[1] The Internet of Things: Making sense of the next mega-trend, Equity Research, The Goldman Sachs Group, Inc., September 3, 2014.

[2] Dave Evans, Cisco Internet Business Solutions Group (IBSG), "The Internet of Things: How the Next Evolution of the Internet Is Changing Everything," April 2011.

[3] Robert C Dixon, *Spread Spectrum Systems with Commercial Applications*. New York, NY: John Wiley & Sons, 1994.

[4] Dr. Jiri Marek – Senior Vice President Research, Bosch, "Emergence of Sensory Swarms for Internet of Things," October 23, 2013.

[5] S.D. Senturia, *Microsystem Design*. Boston, MA: Kluwer, 2001.

[6] H.J. De Los Santos, "Nanoelectromechanical Quantum Circuits and Systems," *Proc. IEEE*, 2003, 91(11): 1907–1921.

[7] Hewlett-Packard Project, "Central Nervous System of the Earth." Available at: https://en.wikipedia.org/wiki/CeNSE.

[8] BI Intelligence, "The Internet of Everything," 2015. Available at: https://www.businessinsider.com/internet-of-everything-2015-bi-2014-12?op=1.

[9] R.P. Feynman, "There's Plenty of Room at the Bottom: An Invitation to Enter a New Field of Physics," American Physical Society Annual Meeting, 29 December, 1959. Available at: http://www.zyvex.com/nanotech/feynman.html.

[10] Héctor J. De Los Santos, *Introduction to Microelectromechanical (MEM) Microwave Systems*. Norwood, MA: Artech House, 1999.

[11] Y.C. Tai, in "MEMS for Optical and RF Applications," University of California at Los Angeles (UCLA) Continuing Education Short Course, 1998 (Co-Instructors H.J. De Los Santos, Y.C. Tai and M.C. Wu).

[12] D.J. Hamilton and W.G. Howard, *Basic Integrated Circuit Engineering*. New York, NY: McGraw-Hill, 1975.

[13] G.E. Moore, "Cramming more components onto integrated circuits," *Electronics*, 1965, 38(8): 56–59.
[14] Available at: https://en.wikipedia.org/wiki/Transistor_count.
[15] Available at: https://en.wikipedia.org/wiki/Physics.
[16] H.J. De Los Santos, "Theory of nano-electron-fluidic logic (NFL): A new digital 'electronics' concept," *IEEE Trans. Nanotechnology*, 2010, 9(4): 474–486.
[17] D. Halliday and R. Resnick, *Physics*, Part II. New York, NY: John Wiley & Sons, 1962.
[18] Héctor J. De Los Santos, *Introduction to Microelectromechanical Microwave Systems*. Norwood, MA: Artech House, 2004.
[19] K.E. Peterson, "Dynamic micromechanics on silicon: Techniques and devices," *IEEE Trans. Electr. Dev.*, 1978, ED-25: 1242–1249.
[20] P.M. Zavracky, S. Majumder, and N. E. McGruer, "Micromechanical switches fabricated using nickel surface micromachining," *J. Microelectromech. Syst.*, 1997, 6: 3–9.
[21] P. Muralt, "MEMS: A Playground for New Thin Film Materials." Available at: http://semiconductors.unaxis.comenchiponline_72dpi issue617-19.pdf.
[22] M.S. Weinberg, "Working equations for piezoelectric actuators and sensors," *ASME/IEEE J. Microelectromech. Syst.*, 1999, 8(4): 529–533.
[23] D. DeVoe, "Thin Film Zinc Oxide Microsensors and Microactuators," Ph.D. Dissertation, Department of Mechanical Engineering, University of California, Berkeley, 1997.
[24] H.-M. Cheng, M.T.S. Ewe, R. Bashir, and G.C.T. Chiu, "Modeling and control of piezoelectric cantilever beam micro-mirror and micro-laser arrays to reduce image banding in electrophotographic processes," *J. Micromech. Microeng.*, 2001, 11: 1–12.
[25] H.B.G. Casimir, "On the attraction between two perfectly conducting plates," *Proc. K. Ned. Akad. Wet.*, 1948, 51: 793.
[26] P.W. Milonni, *The Quantum Vacuum: An Introduction to Quantum Electrodynamics*. San Diego: Academic Press, 1994.
[27] C. Itzykson and J.-B. Zuber, *Quantum Field Theory*. New York, NY: McGraw-Hill International Editions, 1985.
[28] L.S. Brown and G.J. Maclay, "Vacuum stress between conducting plates: An image solution," *Phys. Rev.*, 1969, 184(5): 1272–1279.
[29] I.E. Dzyaloshinskii, "Condensed Matter Physics," Course, Department of Physics, UC Irvine, Spring 2002.

[30] A. Requicha, "Nanorobots, NEMS and nanoassembly," *Proc. IEEE,* 2003, 91(11): 1922–1933.

[31] H.J. De Los Santos, "Impact of the Casimir force on movable-dielectric RF MEMS varactors," *IEEE Nano'03*, August 12–14, San Francisco, CA.

[32] T.H. Boyer, "Van der Waals forces and zero-point energy for dielectric and permeable materials," *Phys. Rev. A*., 1974, 9(5): 2078–2084.

[33] E.Buks and M.L. Roukes, "Stiction, adhesion energy, and the Casimir effect in micromechanical systems," *Phys. Rev. B*., 2001, 63(3): 033402.

[34] T.H. Boyer, "Van der Waals forces and zero-point energy for dielectric and permeable materials," *Phys. Rev*. A, 1974, 9(5): 2078–2084.

[35] O. Kenneth, I. Klich, A. Mann, and M. Revzen, "Repulsive Casimir forces," *Phys. Rev. Lett.*, 2002, 89(3): 033001.

[36] K.A. Milton, The Casimir Effect: Physical Manifestations of Zero-point Energy, World Scientific, Singapore, 2001.

[37] M. Bordag, U. Mohideen, and V.M. Mostepanenko, "New Developments in the Casimir Effect," *Phys. Rep.*, 2001, 353: 1–205.

[38] A. Roy and U. Mohideen, "A verification of quantum field theory—measurement of Casimir force," *Pramana—J. Phys.*, 2001, 56(2&3): 239–243.

[39] B.W. Harris, F. Chen, and U. Mohideen, "Precision measurement of the Casimir force using gold surfaces," *Phys. Rev. A.*, 2000, 62(5): 052109/1-5.

[40] S.K. Lamoreaux, "Experimental verifications of the Casimir attractive force between solid bodies," *Phys. Rev. Lett.*, 1997, 78: 5.

[41] S.K. Lamoreaux, "Demonstration of the Casimir force in the 0.6 to 6 μm range, *Phys. Rev. Lett.*, 1997, 78(1): 5–8.

[42] F. Chen, U. Mohideen, G.L. Klimchitskaya et al., "Demonstration of the lateral Casimir force," *Phys. Rev. Lett.*, 2002, 88(10): 101801/1-4.

[43] G. Bressi, G. Carugno, R. Onofrio et al., "Measurement of the Casimir force between parallel metallic surfaces," *Phys. Rev. Lett.*, 2002, 88(4): 041804.

[44] J. Schwinger, L.L. De Radd, Jt., and K.A. Milton, "Casimir effect in dielectrics," *Ann. Phys.* (NY), 1978, 115: 1.

[45] V.B. Bezerra, G.L. Klimchitskaya, and C. Romero, Casimir force between a flat plate and a spherical lens: Application to the results of a new experiment," *Mod. Phys. Lett.*, 1997, A12: 2613.

[46] N.W. Ashcroft and N.D. Mermin, *Solid State Physics*. Philadelphia: Sounders College, 1976.
[47] E.M. Lifshitz, "The theory of molecular attractive forces between solids," *Sov. Phys. JETP*, 1956, 2: 73.
[48] M. Bordag, U. Mohideen, and V.M. Mostepanenko, "New developments in the Casimir effect," Phys. Rep., 2001, 353(1–3): 1–205.
[49] J. Blocki, J. Randrup, W.J. Swiatecki and C.F. Tsang, "Proximity forces," *Ann. Phys.* (NY) 1977, 105: 427.
[50] E.M. Lifshitz, "The Theory of Molecular Attractive Forces between Solids," *Sov. Phys.*, 1956, 2(1): 73–83.
[51] S.M. Rytov, *Theory of Electrical Fluctuations and Thermal Radiation*, Academy of Sciences USSR, 1953.
[52] I.E. Dzyaloshinskii, E.M. Lifshitz, and L.P. Pitaevskii, "The general theory of Van der Waals forces," *Adv. Phys.*, 1961, 10: 165–209.
[53] L.S. Brown and G. Jordan Maclay, Vacuum stress between conducting plates: An image solution," *Phys. Rev.,* 1969, 184(5): 1272–1279.
[54] T. Emig, N. Graham, R.L. Jaffe, and M. Kardar, "Casimir forces between arbitrary compact objects," *Phys. Rev. Lett.*, 2007, 99: 170403.
[55] K.A. Milton and J. Wagner, "Multiple scattering methods in Casimir calculations," *J. Phys. A*: *Math. Theor.*, 2008, 41: 155402.
[56] M.T. Reid, A.W. Rodriguez, J. White, and S.G. Johnson, "Efficient computation of Casimir interactions between arbitrary 3D objects." *Phys. Rev. Lett.*, 2009, 103(4): 040401.
[57] A.W. Rodriguez, A.P. McCauley, J.D. Joannopoulos, and S.G. Johnson, "Casimir forces in the time domain: I. Theory," *Phys. Rev. A.*, 2009, 80: 012115.
[58] A.P. McCauley, A.W. Rodriguez, J.D. Joannopoulos, and S.G. Johnson, "Casimir forces in the time domain: I. Applications," *Phys. Rev. A.*, 2010, 81: 012119.
[59] J. Babington and S. Scheel, "Casimir forces in multi-sphere configurations," *J. Phys. A: Math. Theor.*, 2010, 43: 215402.
[60] E.M. Lifshitz and L.P. Pitaevskii, *Statistical Physics*: Part 2. Pergamon: Oxford, 1980.
[61] H.J. De Los Santos, "On the status of Casimir force computing for the analysis and design of nanoelectromechanical quantum circuits and systems," NanoMEMS Research, LLC Technical Report NMR-2010-11, 2010.
[62] A. Ashkin, "Acceleration and trapping of particles by radiation pressure," *Phys. Rev. Lett.*, 1970, 24: 156.

[63] D.R. Evans, P. Tayati, H. An, P.K. Lam, V.S.J. Craig, and T.J. Senden, "Laser actuation of cantilevers for picometer amplitude dynamic force microscopy," *Sci. Rep.*, 2014, 4: 5567.
[64] D. Ma, J.L. Garrett, and J.N. Munday, "Quantitative measurement of radiation pressure on a microcantilever in ambient environment," *Appl. Phys. Lett.*, 2015, 106: 091107.
[65] S.R. Rao, "Mirror thermal noise in interferometric gravitational wave detectors," Ph.D. Dissertation, California Institute of Technology, 2003.
[66] H.B. Callen and T.A. Welton, "Irreversibility and generalized noise," *Phys. Rev.*, 1951, 83: 34.
[67] D. Halliday and R. Resnick, *Physics*, Part I. New York, NY: John Wiley & Sons, 1962.
[68] D. Bohm, *Quantum Theory*. New York, NY: Dover Publications, Inc., 1989.
[69] H.-J. Butt and M. Jaschke, "Calculation of thermal noise in atomic force microscopy," *Nanotechnology*, 1995, 6: 1–7.
[70] M. Looney, "An Introduction to MEMS Vibration Monitoring," Analog Dialogue, 2014, 48: 6.
[71] N. Yazdi, F. Ayazi, and K. Najafi, "Micromachined inertial sensors," *Proc. IEEE*, 1998, 86(8): 1640–1659.
[72] H.J. De Los Santos, *Principles and Applications of NanoMEMS Physics*. Dordrecht, The Netherlands: Springer, 2005.
[73] M.A. McCord, A. Dana, and R.F.W. Pease, "The micromechanical tunneling transistor," *J. Micromech. Microeng.*, 1998, 8: 209–212.
[74] H.J. De Los Santos, "A Theoretical Study of Nanoelectromechanical Quantum Tunneling Frequency Multipliers," *German Microwave Conference*, March 16–18, 2009.
[75] H.J. De Los Santos et al., MEMS Tunneling Accelerometer, US8746067, June 10, 2014.
[76] H.J. De Los Santos et al., MEMS Tunneling Accelerometer, US8347720, January 08, 2013.
[77] P. Nelson, "New sound monitoring by IoT can predict mechanical failure," *Network World*, Available at: https://www.networkworld.com/article/2946195/internet-of-things/new-sound-monitoring-by-iot-can-predict-mechanical-failure.html.
[78] Available at: https://en.wikipedia.org/wiki/Fatigue_(material).
[79] Introduction to Shock & Vibration, Available at: https://www.coursehero.com/file/29925789/02-introduction-to-Shock-and-vibration-1pdf/.

[80] H.J. De Los Santos, Christian Sturm, Juan Pontes, *Radio Systems Engineering: A Tutorial Approach*. New York, NY: Springer, 2014.

[81] H.J. De Los Santos, *RF MEMS Circuit Design for Wireless Communications*. Norwood, MA: Artech House, 2002.

[82] H.J. De Los Santos, G. Fischer, H.A.C. Tilmans, J.T.M. van Beek, "RF MEMS for ubiquitous wireless connectivity: Part 1—Fabrication", *IEEE Microwave Magazine*, 2004, 5(4): 36–49.

[83] Stepan Lucyszyn (Editor), *Advanced RF MEMS*, First Edition. Cambridge: Cambridge University Press, 2010.

[84] W. Tian, P. Li, and L. Yuan, "Research and analysis of MEMS switches in different frequency bands," *Micromachines*, 2018, 9: 185.

[85] A. Verger, A. Pothier, C. Guines, A. Crunteanu, P. Blondy, J.-C. Orlianges, J. Dhennin, A. Broue, F. Courtade, and O. Vendier, "Sub-hundred nanosecond electrostatic actuated RF MEMS switched capacitors," *J. Micromech. Microeng.*, 2010, 20: 064011 (7 pp).

[86] Z. Moktadir, S. Boden, H. Mizuta, and H. Rutt, "Graphene for Nano-Electro-Mechanical Systems," *21st Micromechanics and Micro Systems Europe Workshop (MME2010)*, Enschede, The Netherlands September 26–29, 2010.

[87] Z. Shi, H. Lu, L. Zhang, R. Yang, Y. Wang, D. Liu, H. Guo, D. Shi, H. Gao, E. Wang, and G. Zhang, "Studies of graphene-based nanoelectromechanical switches," *Nano Res.*, 2011, 5(22): 82–87.

[88] J.S. Bunch, "Putting a damper on nanoresonators," *Nat. Nanotechnol.*, 2011, 6: 331–332.

[89] S.P. Koenig, N.G. Boddeti, M.L. Dunn, and J.S. Bunch, "Ultrastrong adhesion of graphene membranes," *Nat. Nanotechnol.*, 2011, 6: 543–548.

[90] J.S. Bunch and M.L. Dunn, "Adhesion mechanics of graphene membranes," *Solid State Commun.*, 2012, 152(15): 1359–1364.

[91] P. Sharma, J. Perruisseau Carrier, and A. M. Ionescu, "Nanoelectromechanical Microwave Switch Based on Graphene," *14th IEEE International Conference on Ultimate Integration on Silicon (ULIS)*, March 19–21, 2013.

[92] K.S. Novoselov, A.K. Geim, S.V. Morozov, D. Jiang, Y. Zhang, S.V. Dubonos, I.V. Grigorieva, and A.A. Firsov, "Electric field effect in atomically thin carbon films," *Science*, 2004, 306(5696): 666–669.

[93] G.W Hanson, A.B. Yakovlev, and A. Mafi, "Excitation of discrete and continuous spectrum for a surface conductivity model of graphene," *J. Appl. Phys.*, 2011, 110(11): 114305.

[94] M.F. Craciun, S. Russo, M. Yamamoto et al., "Trilayer graphene is a semimetal with a gate-tunable band overlap," *Nat. Nanotechnol*, 2009, 4(6): 383–388.

[95] Y.-W. Tan, Y. Zhang, K. Bolotin, Y. Zhao, S. Adam, E.H. Hwang, S. Das Sarma, H.L. Stormer, and P. Kim, "Measurement of scattering rate and minimum conductivity in graphene," *Phys. Rev. Lett.*, 2007, 99: 246803.

[96] W. Zhu, V. Perebeinos, M. Freitag, and P. Avouris, "Carrier scattering, mobilities, and electrostatic potential in monolayer, bilayer, and trilayer graphene," *Phys. Rev. B.*, 2009, 80(23): 235402.

[97] S.M. Kim, E.B. Song, S. Lee, S. Seo, D.H. Seo, Y. Hwang, R. Candler, and K.L. Wang, "Suspended few-layer graphene beam electromechanical switch with abrupt on-off characteristics and minimal leakage current," *Appl. Phys. Lett.*, 2011, 99(2): 023103.

[98] S. Bae, H. Kim, Y. Lee, X. Xu, J.-S. Park, Y. Zheng, J. Balakrishnan, T. Lei, H. Ri Kim, Y.I. Song, Y.-J. Kim, K.S. Kim, B. Ozyilmaz, J.-H. Ahn, B.H. Hong, and S. lijima, "Roll-to-roll production of 30-inch graphene films for transparent electrodes," *Nat. Nanotechnol.*, 2010, 5(8): 574–578.

[99] T. Tsang, M. El-Gamal, S. Best, and H.J. De Los Santos, "Wide tuning range RF-MEMS varactors fabricated using the PolyMUMPs foundry," *Microwave J.*, 2003, 46(8): 22–44.

[100] A.A. Generalov, I.V. Anoshkin, M. Erdmanis, D.V. Lioubtchenko, V. Ovchinnikov, A.G. Nasibulin, and A.V. Räisänen, "Carbon nanotube network varactor,' *Nanotechnology*, 2015, 26: 045201 (5 pp).

[101] R.E. Collin, *Foundations of Microwave Engineering*, Second Edition New York, NY: IEEE Press, 2001.

[102] W.-T. Hsu, W.S. Best, H.J. De Los Santos, "Design and fabrication procedure for high Q RF MEMS resonators," *Microwave J.*, 2004.

[103] M. Poot and H.S.J.van der Zant, "Mechanical systems in the quantum regime," *Phys. Rep.*, 2012, 511(5): 273–335.

[104] J.S. Bunch, A.M. van der Zande, S.S. Verbridge, I.W. Frank, D.M. Tanenbaum, J.M. Parpia, H.G. Craighead, and P.L. McEuen, "Electromechanical resonators from graphene sheets" *Science*, 2007, 315: 490–493.

[105] J.W. Jiang, H.S. Park, and T. Rabczuk, "MoS_2 nanoresonators: Intrinsically better than graphene?," *Nanoscale*, 2014, 6(7): 3618–3625.

[106] O. Maillet, X. Zhou, R. Gazizulin, R. Ilic, J. Parpia, O. Bourgeois, A. Fefferman, and E. Collin, "Measuring frequency fluctuations in

nonlinear nanomechanical resonators," *ACS Nano* 2018, 12(6): 5753–5760.

[107] J. Atalaya, "Nonlinear Mechanics of Graphene and Mass-loading Induced Dephasing in Nanoresonators," Ph.D. Dissertation, Department of Applied Physics, Chalmers University of Technology, SE-412 96 Gothenburg, Sweden, 2012.

[108] D. Svintsov, V.G. Leiman, V. Ryzhii, T. Otsuji, and M.S. Shur, "Graphene nanoelectromechanical resonators for the detection of modulated terahertz radiation," *J. Phys. D: Appl. Phys.*, 2014, 47: 505105 (7 pp).

[109] H.C. Nathanson, W.E. Newell, R.A. Wickstrom, and J.R. Davis, "The resonant gate transistor," *IEEE Trans. Electron Devices*, 1967, 14: 117.

[110] Available at: https://en.wikipedia.org/wiki/Ponderomotive_force.

[111] J. Atalaya, A. Isacsson, and J.M. Kinaret, "Continuum elastic modeling of graphene resonators," *Nano. Lett.*, 2008, 8: 4196–4200.

[112] J. Iannacci, "Microsystem based energy harvesting (EH-MEMS): Powering pervasivity of the Internet of Things (IoT) – A review with focus on mechanical vibrations," *J King Saud Univ Sci.*, 2019, 31: 66–74.

[113] S. Kim, R. Vyas, J. Bito, K. Niotaki, A. Collado, A. Georgiadis, and M.M. Tentzeris, "Ambient RF energy-harvesting technologies for self-sustainable standalone wireless sensor platforms," *Proc. IEEE*, 2014, 102(11): 1649–1666.

[114] S. Kitazawa, H. Ban, and K. Kobayashi, "Energy harvesting from ambient RF sources," *Proceedings of the International Microwave Symposium Series – Innovative Wireless Power Transmission (IMWS-IWPT),* Kyoto, Japan, 2012.

[115] M. Piñuela, P.D. Mitcheson, and S. Lucyszyn, "Ambient RF energy harvesting in urban and semi-urban environments," *IEEE Trans. Microwave Theory and Tech.*, 2013, 61(7): 2715–2726.

[116] R. Lu, T. Manzaneque, M. Breen, A. Gao, and S. Gong, "Piezoelectric RF resonant voltage amplifiers for IoT applications," 2016 *IEEE MTT-S International Microwave Symposium (IMS)*, May 22–27, 2016, San Francisco, CA, USA.

[117] C.A. Rosen, *Electromechanical transducer*, US Patent #2830274, 7 March, 1961.

[118] J.M. Gilbert and F. Balouchi, "Comparison of energy harvesting systems for wireless sensor networks," *Int. J. Autom. Comput*, 2008, 5(4): 334–347.

[119] S.J. Roundy, P.K. Wright, and J. Rabaey, "A study of low level vibrations as a power source for wireless sensor nodes," *Comput Commun*, 2003, 26(11): 1131–1144.
[120] S. Saadon and O. Sidek, "Micro-electro-mechanical system (MEMS)-based piezoelectric energy harvester for ambient vibrations" *Procedia Soc Behav Sci.*, 2015, 195: 2353–2362.
[121] J. Iannacci, G. Sordo, M. Schneider, U. Schmid, A. Camarda, and A. Romani, "A novel toggle-type MEMS vibration energy harvester for Internet of Things applications," *Proc. IEEE Sensors*, 2016, 464–466.
[122] S.J. Roundy, "Energy Scavenging for Wireless Sensor Nodes with a Focus on Vibration to Electricity Conversion," Ph.D. Dissertation, University of California, Berkeley, USA, 2003.
[123] S. Boisseau, G. Despesse, and B. Ahmed Seddik, "Electrostatic Conversion for Vibration Energy Harvesting," 2012. Available at: https://www.intechopen.com/books/small-scale-energy-harvesting/electrostatic-conversion-for-vibration-energy-harvesting.
[124] Available at: https://en.wikipedia.org/wiki/Electret.
[125] Y. Huang, R. Doraiswami, M. Osterman, and M. Pecht, "Energy Harvesting Using RF MEMS," 2010 *Electronic Components and Technology Conference*, 1–4 June 2010, Las Vegas, NV, USA.
[126] "Wireless connectivity for the Internet of Things: One size does not fit all," 2017. Available at: http://www.ti.com/lit/wp/swry010a/swry010a.pdf.
[127] "Wireless connectivity for IoT applications." Available at: https://www.mouser.com/pdfdocs/brwireless_web.pdf.
[128] Available at: http://www.technologyuk.net/telecommunications/internet/tcp-ip-stack.shtml.
[129] V.S. Hsu, J.M. Kahn, and K.S.J. Pister, "Wireless communication for Smart Dust," 1998. Available at: http://robotics.eecs.berkeley.edu/~pister/SmartDust/.
[130] Available at: https://iot-analytics.com/10-internet-of-things-applications/.
[131] Available at: https://internetofthingsagenda.techtarget.com/definition/smart-home-or-building.
[132] Available at: http://www.infiniteinformationtechnology.com/iot-wearables-wearable-technology.
[133] Available at: https://www.aglmediagroup.com/smart-cities-and-the-technologies-that-will-transform-our-urban-environments/.

[134] Available at: https://www.scnsoft.com/blog/iot-for-smart-city-use-ca ses-approaches-outcomes.
[135] Available at: https://internet-of-things-innovation.com/industries/.
[136] Available at: https://www.sanog.org/resources/sanog28/SANOG28-Conference_Smart-Grid-with-Internet-of-Things.pdf.
[137] Available at: https://www.networkworld.com/article/3243928/what-is -the-industrial-iot-and-why-the-stakes-are-so-high.html.
[138] Available at: https://www.manufacturingglobal.com/technology/why -your-warehouse-and-factory-operations-need-iiot.
[139] Available at: https://dzone.com/articles/iot-in-transportation-and-log istics-from-fleet-mon.
[140] Available at: https://www.telit.com/industries-solutions/smart-buildin gs/security-surveillance/.
[141] Available at: https://newdeal.blog/how-iot-is-transforming-the-hvacr -industry-26aa24fc3ede.
[142] Available at: https://www.swiftsensors.com/improve-manufacturing-o verall-equipment-efficiency-oee-with-iiot/.
[143] Available at: https://www.telit.com/industries-solutions/agriculture/c rop-livestock-monitoring/.
[144] Available at: https://en.wikipedia.org/wiki/Connected_car.
[145] Available at: https://igniteoutsourcing.com/automotive/connected-car -and-iot/.
[146] Available at: https://www.businessinsider.com/internet-of-things-con nected-smart-cars-2016-10?op=1.
[147] Available at: https://www.intellimec.com/ims-blog/connected-car-an alytics.
[148] Available at: http://minnova.am/iot-applications.
[149] Available at: https://iotworm.com/connected-healthcare-internet-thin gs-iot-examples/.
[150] Available at: https://www.iotforall.com/retail-iot-applications-challe nges-solutions/.
[151] Available at: https://global-sensors.com/temperature-monitoring-blog /let-our-temperature-monitors-ensure-your-food-safety/.
[152] Available at: https://www.blumeglobal.com/learning/internet-of-things/.
[153] Available at: https://easternpeak.com/blog/iot-in-agriculture-5-techno logy-use-cases-for-smart-farming-and-4-challenges-to-consider/.

[154] R. Thirunarayanan, A. Heragu, D. Ruffieux, and C. Enz, "Ultra low-power MEMS based radios for the IoT," *International Conference on Electronics and Communication Systems (ICECS)*, 2016, 225–228.
[155] S. Samadi and H. Mirzaee, "Wireless Sensor Technologies in Food Industry: Applications and Trends," *INNOV 2015: The Fourth International Conference on Communications, Computation, Networks and Technologies*, 2015, 74–78.
[156] J.G. Andrews, S. Buzzi, W. Choi et al., What will 5G be?, *IEEE J. on Selected Areas in Comm.*, 2014, 32(6): 1065–1082.
[157] H.J. De Los Santos, C. Sturm, and J. Pontes, *Radio Systems Engineering: A Tutorial Approach*. Berlin: Springer, 2014.
[158] D. Choudhury, "5G Wireless and Millimeter Wave Technology Evolution: An Overview," 2015 *IEEE MTT-S International Microwave Symposium*, 17–22 May 2015, Phoenix, AZ, USA.
[159] T.S. Rappaport, S. Sun, R. Mayzus, H. Zhao, Y, Azar, K. Wang, G.N. Wong, J.K. Schulz, M. Samimi, and F. Gutierrez, "Millimeter Wave Mobile Communications for 5G Cellular: It Will Work!," *IEEE Access*, 2013, 1: 335–349.
[160] L. Wei, R.Q. Hu, Y. Qian, and G. Wu, "Key Elements to Enable Millimeter Wave Communications for 5G Wireless Systems," *IEEE Commun Mag.*, 2014, 137–143.
[161] R.C. Jaeger, *Introduction to Microelectronics Fabrication*, Volume V, Modular Series on Solid State Devices, G.W. Neudeck and R.F. Pierret (Eds.), Addison-Wesley, 1988.
[162] W. Bacher, W. Menz, and J. Mohr, "The LIGA technique and its potential for microsystems—A survey," *IEEE Trans. Ind. Electronics*, 1995, 42(5): 431–441.
[163] J. Bryzek, K. Petersen, and W. McCulley, "Micromachines on the march," *IEEE Spectrum*, 1994, 31(5): 20–31.
[164] H.C. Nathanson, W.E. Newell, R.A. Wickstrom, and J.R. Davis, Jr., "The resonant gate transistor," *IEEE Trans. Electron Dev.*, 1967, 14(3): 117–133.
[165] R.T. Howe and R.S. Muller, "Polycrystalline silicon micrimechanical beams," *J. Electrochemical Soc.*, 1983, 130: 1420–1423.
[166] M. Mehregany, K.J. Gabriel, and W.S.N. Trimmer, "Integrated fabrication of polysilicon mechanisms," *IEEE Trans. Electron Dev.*, 1988, 35(6): 719–723.

[167] L.-S. Fan, Y.-C. Tai, and R.S. Muller, "Integrated movable micromechanical structures for sensors and actuators," *IEEE Trans. Electron Dev.*, 1988, 35(6): 724–730.

[168] M. Parameswaran, H.P. Baltes, and A.M. Robinson, "Polysilicon Microbridge Fabrication Using Standard CMOS Technology," Tech. Digest, *IEEE Solid-State Sensor and Actuator Workshop*, 1988, 148–150.

[169] M.W. Putty, S.C. Chang, R.T. Howe, A.L. Robinson, and K.D. Wise, "Process integration for active polysilicon resonant microstructures," *Sensors & Actuators*, 1989, 20: 143–151.

[170] L.E. Larson, R.H. Hackett, and R.F. Lohr, "Microactuators for GaAs-Based Microwave Integrated Circuits," *IEEE Conference on Solid State Sensors and Actuators*, 1991, 743–746.

[171] H.J. De Los Santos et al., Microelectromechanical Device, US6040611, March 21, 2000.

[172] K.R. Williams and R.S. Muller, "Etch rates for micromachining processing," *J. Microelectromech. Syst.*, 1996, 5(4): 256–259.

[173] M.A. Schmidt, R.T. Howe, S.D. Senturia et al., "Design and calibration of a microfabricated floating-element shear-stress sensor," *IEEE Trans. Electron Dev.*, 1988, 35(5): 750–757.

[174] O. Tabata, H. Funabashi, K. Shimoaka et al., "Surface micromachining using polysilicon sacrificial layer," *The Second International Symposium on Micromachine and Human Science*, Japan, 1991.

[175] A.B. Frazier and M.G. Allen, "High aspect ratio electroplated microstructures using a photosensitive polyimide process," *Proceedings of the IEEE MEMS 92*, February 1992, 87–92.

[176] A. Nathan, A. Ahnood, M.T. Cole, S. Lee, Y. Suzuki, P. Hiralal, F. Bonaccorso, T. Hasan, L. Garcia-Gancedo, A. Dyadyusha, S. Haque, P. Andrew, S. Hofmann, J. Moultrie, D. Chu, A.J. Flewitt, A.C. Ferrari, M.J. Kelly, J. Robertson, G.A. J. Amaratunga, and W.I. Milne, "Flexible electronics: The next ubiquitous platform," *Proc. IEEE*, 2012, 100: 1486–1517.

[177] S.P. Li, D.P. Chu, C.J. Newsome, D.M. Russell, T. Kugler, M. Ishida, and T. Shimoda, "Short-channel polymer field-effect-transistor fabrication using spin-coating-induced edge template and ink-jet printing," *Appl. Phys. Lett.*, 2005, 87(23): 232111.

[178] S.M. Ferreira Cruz, L.A. Rocha, and J.C. Viana, "Printing Technologies on Flexible Substrates for Printed Electronics, Flexible

Electronics, Simas Rackauskas, IntechOpen," 2018. doi: 10.5772/intechopen.76161. Available at: https://www.intechopen.com/books/flexible-electronics/printing-technologies-on-flexible-substrates-for-printed-electronics.
[179] Available at: https://en.wikipedia.org/wiki/Screen_printing.
[180] Available at: https://www.lifewire.com/flexography-printing-technique-1074610.
[181] Available at: https://www.lifewire.com/what-is-gravure-printing-1074611.

Index

About the Author

Héctor J. De Los Santos received the Ph.D. degree in electrical engineering from Purdue University, West Lafayette, IN, in 1989. He is currently President & CTO of NanoMEMS Research, LLC, a company he founded in 2002. From 2000 to 2002 he was Principal Scientist at Coventor, Inc., Irvine, CA. From 1989 to 2000, he was with Hughes Space and Communications Company, Los Angeles, CA, where he served as Principal Investigator and the Director of the Future Enabling Technologies IR&D Program. He is author of five books and holds over 30 U.S., European, German and Japanese patents. His research interests include, theory, modeling, simulation, and design of emerging devices (electronic, plasmonic, nanophotonic, nanoelectromechanical, etc.), and wireless communications. During the 2010–2011 academic year, he held a German Research Foundation (DFG) *Mercator Visiting Professorship* at the Institute for High-Frequency Engineering and Electronics, Karlsruhe Institute of Technology, Germany. He is an IEEE Fellow.